Geometric Dimensioning and Tolerancing
Applications and Inspection

Second Edition

Gary K. Griffith

Upper Saddle River, New Jersey
Columbus, Ohio

Library of Congress Cataloging-in-Publication Data

Griffith, Gary K.
   Geometric dimensioning and tolerancing / Gary K. Griffith.—2nd ed.
     p. cm.
   Originally published: Measuring and gaging geometric tolerances. Englewood Cliffs,
N.J.: Prentice Hall Career and Technology, ©1994.
   Includes index.
   ISBN 0-13-060463-1
   1. Tolerance (Engineering) I. Title.
TS172 .G74 2002
620´.0045—dc21   2001021999

Editor in Chief: Stephen Helba
Executive Editor: Debbie Yarnell
Production Editor: Louise N. Sette
Production Supervision: Carlisle Publishers Services
Design Coordinator: Robin G. Chukes
Cover Designer: Mark Shumaker
Production Manager: Brian Fox
Marketing Manager: Jimmy Stephens

This book was set in 10.5/12 Times by Carlisle Communications, Ltd. It was printed and bound by R. R. Donnelley & Sons Company. The cover was printed by LeHigh Press, Inc.

First edition, © 1994, entitled *Measuring and Gaging Geometric Tolerances.*

Prentice-Hall International (UK) Limited, *London*
Prentice-Hall of Australia Pty. Limited, *Sydney*
Prentice-Hall Canada, Inc., *Toronto*
Prentice-Hall Hispanoamericana, S.A., *Mexico*
Prentice-Hall of India Private Limited, *New Delhi*
Prentice-Hall of Japan, Inc., *Tokyo*
Prentice-Hall Singapore Pte. Ltd.
Editora Prentice-Hall do Brasil, Ltda., *Rio de Janeiro*

Copyright © 2002, 1994, by Pearson Education, Inc., Upper Saddle River, New Jersey 07458.
All rights reserved. Printed in the United States of America. This publication is protected by Copyright and permission should be obtained from the publisher prior to any prohibited reproduction, storage in a retrieval system, or transmission in any form or by any means, electronic, mechanical, photocopying, recording, or likewise. For information regarding permission(s), write to: Rights and Permissions Department.

10 9 8 7 6 5 4 3 2 1
ISBN 0-13-060463-1

To Sharon, Christopher, Alan, Brian, and Kimberlee

Contents

Preface xvii

1 Introduction   1

Objectives   1
General Symbols and Pertinent Definitions   2
  Geometric Tolerancing Symbols   3
  Tolerance Zones   4
  Tolerance Zone Modifiers   5
General Rules for Geometric Tolerancing   7
Free-State Variation   10
Restrained Features   11
Measurement Accuracy and Precision   11
Review Questions   12

2 Datums   13

Objectives   13
Key Facts about Datums   13

Functional and Nonfunctional Datums   14
Simulated Datums   16
Qualified Datums   17
Contacting Functional Datums   18
Compound Datums   19
Offset Compound Datums   20
Equalizing Datums   20
Datum Targets (Nonfunctional Datums)   21
Review Questions   23

3 Inspecting Size Tolerances   24

Objectives   24
Introduction   24
Noncylindrical Parts   25
Cylindrical Features: Shafts   28
Cylindrical Features: Holes   28
Problems with Gages   32
Alternatives to Hard Gages   32
Summary   35
Review Questions   35

4 Flatness   36

Objectives   36
Introduction and Applications   36
Key Facts   37
Flatness Measurement: Jackscrew Method   38
Flatness Measurement: Wobble-Plate Method   41
Flatness Measurement: Fixed-Plane Methods   41
Flatness Measurement: Direct-Contact Method   43
Optical Flatness Inspection   45
    Interference Light Bands (Fringes)   45
    Procedure for Testing Flatness with Optical Flats   45

Wedge Method   46
Contact Method   46
If Bands Do Not Appear   46
Interpreting the Bands for Flatness Error   47
Review Questions   49

5 Straightness   50

Objectives   50
Introduction and Applications   50
Key Facts: Straightness of Surface Elements   52
Key Facts: Straightness of an Axis   53
Key Facts: Straightness of a Centerplane   53
Straightness of Surface Elements: Cylindrical Parts   53
    Straightness of Surface Elements: Jackscrew Method   53
    Straightness of Surface Elements: Precision Straightedge Method   57
    Straightness of Surface Elements: Comparator Method   58
    Straightness of Surface Elements: Two-Block Method   58
Straightness of Surface Elements: Noncylindrical Parts   59
    Applications   60
Straightness of an Axis   60
Straightness of a Centerplane   63
Introduction to Functional Gaging: Straightness Examples   64
Straightness of an Axis: RFS   66
    Straightness of an Axis: Differential Measurement Method   66
    Straightness of an Axis: Precision Spindle Method   68
Straightness of a Centerplane at MMC   70
    Functional Gaging Straightness of a Centerplane at MMC   72
    Simulating a Functional Gage for Straightness of Centerplane at MMC   74
Straightness of a Centerplane (RFS)   75
    Straightness of a Centerplane: Differential Measurement Method   76
Review Questions   78

6 Circularity 79

Objectives 79
Introduction and Applications 79
Key Facts 80
The Problem of Lobes 81
The Effective Size 81
Vee-Block Method: Outside Circularity Measurement 83
Vee-Anvil Micrometer: Outside Circularity Measurement 87
Bore Gage Method: Inside Circularity Measurement 88
Verifying Circularity Using Runout to Centers 89
Rotary Table Method 89
Pneumatic Gages for Circularity Measurement 91
Precision Spindle Method 91
  Polar Graphs 94
Summary 94
Review Questions 95

7 Cylindricity 96

Objectives 96
Introduction and Applications 96
Key Facts 98
Open-Setup Method 98
Cylindricity Verified Using Total Runout 98
Precision Spindle Method 100
Review Questions 102

8 Parallelism 103

Objectives 103
Introduction and Applications 103
Key Facts 104
Surface-to-Surface Parallelism: Surface-Plate Method 105
Axis-to-Surface Parallelism: Surface-Plate Method 108

Axis-to-Surface Parallelism: Bonus Tolerances   111
Axis-to-Axis Parallelism: Surface-Plate Method   112
Parallelism: Secondary Alignment Datums   116
Parallelism: Compound Datums   118
Review Questions   122

9 Perpendicularity   123

Objectives   123
Introduction and Applications   123
Key Facts   124
Implied Perpendicularity   124
Perpendicularity: Single Surface, One Datum   124
Perpendicularity: Precision-Square Method   126
Perpendicularity: Cylindrical-Square Method   126
Special Indicator Stands for Perpendicularity Inspection   129
Perpendicularity: Right-Angle-Plate Method   129
Vertical Measuring Systems   134
Perpendicularity of an Axis (RFS)   135
Perpendicularity of an Axis (MMC)   139
Perpendicularity of an Axis to an Axis   140
Review Questions   145

10 Angularity   146

Objectives   146
Introduction and Applications   146
Key Facts   147
Surface-to-Surface Angularity   147
Measuring Angularity with a Sine Bar or Sine Plate   149
Sine-Bar Method: Angularity with Secondary Datums   152
Sine Plates   154
Angular-Gage-Block Method   154
Angularity of a Size Feature   156

Angularity of a Hole: Sine-Bar Method   156
Angularity of a Cone   160
Setting Compound Sine Plates   164
Upside-Down Use of the Sine Bar   165
Review Questions   166

11 Circular Runout   168

Objectives   168
Introduction and Applications   168
Key Facts   169
Circular Runout: Vee-Block Method   169
Circular Runout: Inside-Diameter Datums   172
End-Surface Circular Runout: Two Datums   174
Circular Runout Applied to a Cone   176
Circular Runout: Compound Offset Datums   179
Review Questions   181

12 Total Runout   183

Objectives   183
Introduction and Applications   183
Key Facts   184
Total Runout: Outside-Diameter Datum   184
Total Runout: Inside-Diameter Datum   186
Total Runout: End Surface   189
Review Questions   192

13 Profile of a Line   193

Objectives   193
Introduction and Applications   193
Key Facts   194
General Measurement of Profile Tolerances   194

Profile Dimensioning and Tolerance Zones   195
Profile of a Line: No Datums   195
Optical Comparator Measurement   197
Profile of a Line Using Datums   201
Mechanical Gaging: Profile Tolerances   203
Profile: To Locate a Surface   204
Review Questions   205

14   Profile of a Surface   207

Objectives   207
Introduction and Applications   207
Key Facts   208
Profile of a Surface: No Datums   208
Profile of a Surface Using Datums   210
Profile for Coplanar Surfaces: Introduction   211
Profile for Coplanar Surfaces: Datums Applied   211
Profile for Coplanar Surfaces: Datums Not Applied   215
Quick-Check Method: Height Differences   215
Coplanarity Measurement: Jackscrew Method   218
Coplanarity: Compound Datums   218
Review Questions   219

15   Concentricity   221

Objectives   221
Introduction and Applications   221
Key Facts   222
Difference Between Concentricity and Runout   222
Differential Measurements   223
Concentricity Verified by Total Runout   227
Precision Spindle Method   227
Review Questions   229

16 Position Tolerances 231

Objectives 231
Introduction and Applications 231
Key Facts 232
Coordinate (±) Dimensioning Versus Position Tolerancing 232
Conversion of Cylindrical Zones to Coordinate (±) Tolerances 236
Conversion of Coordinates to Cylindrical Zones 236
Bonus Tolerances 237
Single Feature Position Tolerance 238
Coaxial Features 243
Position: Coaxial at MMC 244
Feature Pattern Locations 247
Surface-Plate Setup 250
    Simple Caliper Measurement 255
    Functional Gaging at MMC 255
Feature-to-Feature and Perpendicularity Control 258
Pattern Location, Feature to Feature, and Perpendicularity 263
Position Tolerances: Cylindrical Parts, Tertiary Datum 267
Composite Position Tolerances 271
Bidirectional Position Tolerances 274
Projected Tolerance Zones 278
Zero Position Tolerance at MMC 282
Position Tolerance at LMC 284
    Position: Two Single Segment 285
Boundary Position Tolerancing 287
    Position Tolerance: Separate Requirements 288
    Position Tolerance: Conical Tolerance Zone 290
    Position Tolerance: At LMC 291
Coordinate Measuring Machines 293
    Summary 298
Review Questions 298

Contents **xiii**

17 Symmetry   299

Introduction and Applications   299
Key Facts   299
An Example on a Drawing   300
Review Questions   303

18 Introduction to Functional Gage Design   304

Introduction   304
Functional Gage Design Principles   304
Functional Gage Advantages and Disadvantages   304
  Advantages of Functional Gages   305
  Disadvantages of Functional Gages   306
Functional Gaging Costs   306
Basis for Decisions to Use Functional Gages   307
Materials Used for Functional Gages   307
  Carbon Steels   308
  Tool Steels   308
  Chrome-Plated Gaging Members   308
  Carbide Steels   309
  Ceramics   309
Surface Roughness (Texture)   309
Gage Makers' Tolerance and Wear Allowance   309
  Establishing Gage Wear, Allowances   310
Geometric Tolerances That Cannot Be "Gaged"   310
  The Effect of "Tight" Tolerances on Gage Designs   311
Part Datums and Functional Gage Interfaces   311
Functional Gage Interfaces for Size Feature Datums   312
Size Feature Datum–Virtual Condition Rule   312
Gage Pins   314
Fixed Pins Versus Sliding Pins   314
Functional Gage Design Examples   315
  Perpendicularity of a Hole at MMC   315

  Straightness of an Axis at MMC 316
  Position of a Hole to Three Datum Planes 316
  Coaxial Position Tolerance 319
  Coaxial Position Tolerance and Pattern Location 320
  Position Tolerance—Intrinsic Datums 321
  Position Tolerance and Pattern Location 322
Alternatives to Functional Gaging 322
Using Functional Gages 324
Examples of the Three Actual Gage Designs 325
  Part 1: Position Tolerance—MMC Size Datum 325
  Part 2: Position Tolerance—Virtual Condition Datum 327
  Part 3: Position of a Two-Hole Pattern to a Primary Datum Face 329
Coordinate Measuring Machines (CMMs) 333
Summary 335
Review Questions 335

Appendices

A Conversion of Actual Coordinate Measurements to Position Location Diameter 337
B Conversion from Diametral Tolerance Zone to Plus and Minus Coordinate Tolerances 338
C Bolt Circle Chart 339
D Differences Between ASME and ISO Standard Symbols 340
E Other Drawing Symbols 341
F Table of Trigonometric Functions 342

Glossary 343

Solutions to Odd-Numbered Chapter Review Questions 345

Index 347

Preface

A variety of publications cover the American Society for Mechanical Engineers (ASME) Y14.5M standard Geometric Dimensioning and Tolerancing (GDT). One problem faced by those who are responsible for inspection is the fact that most of these publications cover the subject of GDT only from the point of view of the draftsperson, designer, or manufacturing engineer. There are few sources of information for the people who have to inspect the end product with respect to these tolerances. Many who understand geometric tolerances and inspection methods have learned them the hard way—through mistakes and experience. The primary purpose of this book is to provide a source of learning for specific methods and techniques that can be used to ensure functional inspection of geometric tolerances. This book is designed specifically for operators, machinists, inspectors, quality technicians and engineers, and other people in manufacturing companies who have the need to know about inspection methods and techniques.

Some assumptions were made during the preparation of this book. One is that the reader already understands how to use and care for standard inspection equipment. A variety of inspection gages and measuring instruments (such as micrometers, indicators, surface plates, right-angle plates, and precision parallels) will be used in the examples, and it is assumed that the reader knows how to use this equipment. Many available textbooks cover the fundamentals of measuring and gaging equipment if the reader should need to study these topics.

Another assumption is that the reader works with GDT on engineering drawings per the ASME Y14.5M standard. The ASME symbols are interpreted in this book to the extent necessary to understand the inspection or gaging method discussed. An introduction to each symbol and some *key facts* about the symbol provide the reader with the interpretation necessary to perform the inspection and to better understand the symbol and its application. There are many different applications for GDT on a wide variety of products and functional requirements. It is not possible to cover all the different combinations and applications in one text. This book does, however, cover all the geometric tolerances in the standard, a representative sample of applications, and more than one method of inspection for each symbol.

Coverage of inspection methods for each symbol will begin with a sample drawing of a part, an explanation of the geometric requirement, a review of the GDT rules and tolerance zones that apply, and, finally, step-by-step instructions and photos for the inspection or gaging method. Readers should try to *consider an application of their own* as they are studying the various inspection and gaging methods covered in this book. For example, the methods illustrated might cover an example using a small part that are the same methods for larger parts using larger equipment. In every case, the rules and tolerance zones for each symbol are the same as the examples indicate, except that different amounts of tolerances are often applied.

It is important to understand that all the tolerance values used in this book are examples only and do not in any way constitute standard tolerance values. Tolerances for a given design depend solely on the design requirement. All inspection methods and techniques carry with them a certain amount of inherent error. The limitations of each method will also be covered so that appreciable measurement errors can be avoided.

Another assumption is that readers will use the knowledge obtained from each application covered to help them see how the inspection could be performed with other similar instruments, keeping in mind that a wide variety of measuring machines could be used for the inspection.

Most of the parts used in this book, except those with a specific credit line, have been machined specifically for use in the book and are the property of the author. All the measuring instruments used in this book are *used* instruments that have been donated by various companies, or the company has allowed the author access to its equipment. Although most of the instruments are second-hand, they provide good physical examples of each measurement method covered. This second edition offers the following changes and additions:

- Extensive revisions have been made to all text and graphics to update the book to the latest revision of the reference standard (American Society for Mechanical Engineers (ASME) Y14.5M—1994)).
- Symmetry has been separated into a new Chapter 17.
- A new Chapter 18 has been added, entitled Introduction to Functional Gage Design.
- New photos of functional gages have been added in Chapter 18.
- New applications have been covered that are found only in the latest 1994 ASME standard.

Finally, this book has been limited by the type and amount of gages and products that have been donated for use in the examples. There are many different kinds of equipment on the market today that are more sophisticated than the equipment used in this book, but the principles of measuring geometric tolerances are well demonstrated by these traditional measuring tools and gages.

ACKNOWLEDGMENTS

I would like to thank the following companies and individuals who provided support and equipment or products for completing this project. A special thanks goes to Garrett Automotive Products Company, Torrance, California (specifically to Andy Timko),

Preface **xvii**

who provided many of the measuring tools and parts that were used in this project, and to AlliedSignal Aerospace Systems & Equipment, Torrance, California (specifically to Angela Elliott and Victor Martinez), who also provided many of the measuring tools.

A special thanks also goes to Woodruff Corporation, Torrance, California; A. H. Machine Inc., Inglewood, California; Delco Machine & Gear, Gardena, California; Barry Controls Aerospace, Burbank, California; and Ace Air Inc., Gardena, California, who provided special parts for use in the book, and to James Pluma, Corona, California; who took all the photographs of inspection setups.

I would also like to acknowledge and thank the reviewers who provided valuable feedback and helpful suggestions in preparation of this revision. Those reviewers include: Thomas G. Soyster, Eastern Michigan University and Calvin, J. Sellers, San Joe City College.

PERMISSIONS

Due to the numerous equipment manufacturers whose products are in some cases shown in a single photograph, we would like to acknowledge here these manufacturers and products, thus avoiding extensive credit lines under each photo. The courtesy shown by these companies is greatly appreciated.

Measuring Equipment	Courtesy of
Dial indicator, Intra-Mic	Brown & Sharpe, Manufacturing Corporation, North Kingstown, Rhode Island
Micrometers, telescoping gages, adjustable parallels, dial indicator, matched vee blocks, drop indicator, Webber gage blocks, height gages	L. S. Starrett Company, Athol, Massachusetts
Roundness measuring machine, ring gages	Federal Products Corporation, Providence, Rhode Island
Dial caliper	Fred V. Fowler Co., Inc., Newton, Massachusetts
Surface gage	Mitutoyo/MTI Corporation, Aurora, Illinois
Gage pins	Deltronic Corporation, Santa Ana, California
Matched precision parallels	Taft-Peirce, Woon, Rhode Island
Granite surface plate	Continental Granite Corporation, Escondido, California
Clamps	Clamp Manufacturing Company, Inc., South El Monte, California

1

Introduction

OBJECTIVES

Upon completion of this chapter, the reader should be able to:

1. Understand general symbols and definitions.
2. List the geometric tolerancing symbols and the families to which each symbol belongs.
3. Understand tolerance zone modifiers, maximum material condition (MMC), and virtual condition.
4. Understand and apply the five general rules for geometric tolerancing.
5. Define how measurement accuracy and precision are obtained.

Unlike simple dimensions, geometric tolerances, when properly applied to product designs, are always functional or have a direct bearing on the fit, form, and/or function of the part. Product designers count on these geometric controls to provide design assurance that the product will have the appropriate fit, form, function, and/or reliability. Geometric tolerancing symbols are used to specify the functional relationship between or within part features and have specific rules about them that have a direct bearing on how the part must be inspected.

Unlike simple dimensioning, a geometric tolerance defines *datums* that are the starting point for the measurement and a specific description of the *tolerance zone* so that the observer can decide if the product is within or out of specification. These datums and tolerance zones, to a large degree, set the ground rules for appropriate inspection methods and techniques.

The *observer* (or the person making the measurement) has the tasks of (1) understanding geometric tolerancing symbols and the rules about them, (2) understanding a variety of inspection and gaging equipment, and (3) finding the appropriate measuring or gaging method that will properly evaluate the part feature being controlled.

It is important to understand that the setup for measurement, or the design of a gage for geometric tolerances, must *properly contact datum references and be able to reproduce the specific tolerance zone for the part, or it cannot be used to measure the geometric requirement.* This is one of the fundamental lessons to be learned in measuring or gaging geometric tolerances. Remembering this basic principle will help the observer perform proper and functional inspection. The following pages in this chapter are devoted to covering the principles and rules of geometric tolerancing as they relate to inspection methods. These principles will then be applied throughout the rest of the book to reinforce learning.

GENERAL SYMBOLS AND PERTINENT DEFINITIONS

A variety of symbols and definitions are used in geometric tolerancing to replace wordy notes on the drawing. Each symbol has a specific and consistent meaning. Refer to the glossary for further definitions.

Features: Physical portions of a part, such as surfaces, holes, tabs, bosses, contours, screw threads, or slots.

Size features: Dimensions of size (for example, for holes, slots, tabs, or bosses).

Feature control frame: Contains the geometric symbol, the geometric tolerance, and specifies all applicable datums.

Datums: Imaginary planes, axes, points, lines, or arcs that are derived by contacting datum features. Datum features are physical features of the part that are used to establish datums.

Datum targets: Specific points, lines, or areas that are most widely used to establish datums for making the part. These datums are often referred to as *construction* datums and are generally not permanent.

Tolerance zones: All geometric tolerances have imaginary tolerance zones that are the basis for acceptance or rejection of the product. These tolerance zones have specific shapes, depending on the geometric tolerance and the feature being controlled.

Bonus tolerances: When tolerances are modified with MMC or LMC modifiers, a bonus tolerance is allowed when the actual size of a feature being controlled departs from the size boundary indicated by the modifier.

Functional datum symbol: Any letter of the alphabet (except I, O, or Q) that is shown in a box with a hyphen on the left and right. See Figure 1.1.

Datum target symbol: There are three kinds of datum targets and two symbols that identify them. Datum target lines and points are identified with one symbol, and datum target areas are identified with another. Refer to Figure 1.1.

Diameter symbol: The diameter symbol is a circle with a slash through it at a 45° angle, as shown in Figure 1.1. The diameter symbol precedes all diameters on drawings and is also used to symbolize a cylindrical tolerance zone.

General Symbols and Pertinent Definitions   3

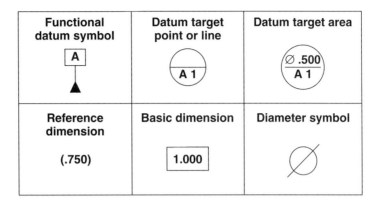

Figure 1.1 ASME Y14.5 symbols.

Basic dimension: A numerical value that is used to describe the theoretically exact size, profile, orientation, or location of a feature. Basic dimensions have no tolerance and cannot be rejected. A basic dimension is shown by the word *basic* or the abbreviation *bsc* after the dimension, and the preferred method is to show the dimension in a box (see Figure 1.1).

Reference dimensions: Reference dimensions are for computational or informational purposes only. They are not inspected and therefore cannot be rejected. A reference dimension is shown with the abbreviation *ref* after it; the preferred method is to show it in parentheses (see Figure 1.1).

Full indicator movement (FIM) (formerly *total indicator reading (TIR)*: The total travel of a dial indicator pointer during a measurement. For example, when you are using a .001″ balanced dial indicator, if the pointer travels as low as minus .002″ to as high as plus .003″, the FIM is .005″.

Geometric Tolerancing Symbols

Five families of geometric control symbols will be covered in this book. Each family of symbols is used for specific controls. The following is an introduction to each family.

Form tolerances: Form tolerances are designed to control the form (or shape) of individual features and features of size. The form tolerance family includes *flatness, straightness, circularity (roundness),* and *cylindricity.* Form tolerances control individual features and do not control the relationship of one feature to another.

Orientation tolerances: Orientation tolerances control specific relationships of one feature to another; therefore, they are always specified to at least one datum reference. The orientation tolerance family includes *parallelism, perpendicularity,* and *angularity.*

Runout tolerances: Runout tolerances apply to rotating parts in order to control the coaxiality of cylindrical features to one another or the runout of end surfaces with respect to datum axes. The runout family includes *circular runout* and *total runout*.

Profile tolerances: Profile tolerances are used to control irregular shapes such as contours and can also be applied to control coplanarity (more than one surface in the same plane). The profile family includes *profile of a line* and *profile of a surface*.

Location tolerances: Location tolerances are used to control the location of the center of size features (such as the location of the axis of a hole or pin or the centerplane of a slot or square boss). The location tolerance family includes *position, symmetry,* and *concentricity*.

Tolerance Zones

All geometric tolerances have specific imaginary tolerance zones (see Figure 1.2) that tend to take the shape of the feature being controlled. For example, positioning a hole (which is a cylinder) is a *cylindrical tolerance zone* that locates the axis of the hole, and concentricity is also a *cylindrical tolerance zone.* Parallelism of a surface tends to have a tolerance zone that will surround a surface *(two imaginary parallel planes that are perfectly parallel to the datum). Two imaginary planes* also define the tolerance zone for flatness of a surface; but since flatness has no datum, the two planes are independent.

The tolerance zone for straightness of the surface of a pin is *two straight lines,* yet the tolerance zone for straightness of the axis of the pin is a *cylinder.* It is important for the observer to understand geometric tolerance zones so that functional inspection can result. Specific tolerance zones help identify the proper gaging and technique because the observer must prove, through inspection, that the controlled feature is either within or outside the zone of tolerance.

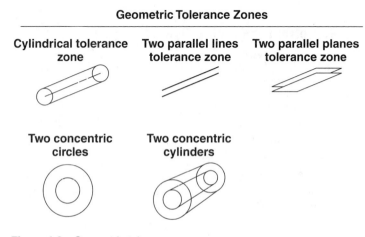

Figure 1.2 Geometric tolerance zones.

Tolerance Zone Modifiers

Certain modifiers used in geometric tolerancing can alter the stated tolerance zone when they are specified. Although the modifiers use specific symbols, let's first discuss their meaning:

Maximum material condition (MMC): The maximum material condition applies to all size features. It is the condition of size (within size tolerances) for which the size feature has the maximum material allowed (such as the largest shaft size or the smallest hole size). When MMC is called out, it means that the stated geometric tolerance applies only when the controlled feature is at MMC size.

Least material condition (LMC): The least material condition also applies to size features. It is the condition of size (within size tolerances) for which the size feature has had all the material removed that is allowed to be removed (for example, the smallest allowable shaft size or the largest allowable hole size). When LMC is called out, it means that the stated geometric tolerance applies only when the part is at LMC size.

Regardless of feature size (RFS): This modifier indicates that the stared geometric tolerance value is required no matter what size the feature is (within its size tolerances). RFS nullifies the benefits of MMC and LMC.

Projected tolerance zone: This modifier extends a tolerance zone out in space beyond the feature being controlled. (For further discussion on this modifier, refer to page 278.

Free state: This modifier replaces the words "free state" that used to be stated under the feature control frame. It applies to parts that are subject to free-state variation. (Refer to Figure 1.6 later in the chapter for an example.)

Tangent plane: This modifier is used when the designer is only concerned about controlling the high points of a surface (or plane). It means that a precision plane must contact the surface being controlled and measurements are made on the plane itself. (Refer to Figure 1.7 later in the chapter.)

Figure 1.3 shows the symbols for these modifiers. The modifiers and their application will be discussed in detail in the following pages as each geometric symbol is discussed and inspected.

The Concept of MMC and Virtual Condition In many design applications, especially for mating parts, MMC principles are used to obtain guaranteed fits without the necessity to tighten tolerances. In fact, using MMC principles, required fits are obtained while, at times, allowing more tolerance for the parts. MMC is the maximum boundary of the size limit for outside size features (such as shafts, pins, and bosses) *and* is the minimum size boundary of the size limit for inside features (such as holes and slots).

A simple example is an assembly where a shaft must go into a hole. The only function is that the shaft go through the hole with a *minimum* clearance of .002″. In this case, if the tolerance limits on the shaft and the hole were set to .002″ clearance between their

6 Chapter 1 Introduction

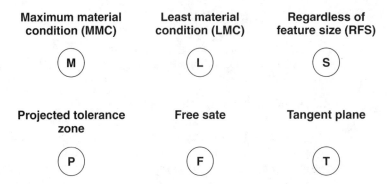

Figure 1.3 Material condition modifiers and projected tolerance zone modifier.

respective MMC limits, this fit would be guaranteed as long as the shaft does not exceed its MMC limit (oversize) and the hole does not exceed its MMC limit (undersize). Hence, at the MMC limits of both mating parts, there is .002″ clearance.

MMC boundaries control the effective size of size features, but do not control their location or orientation to other features. Using the MMC boundaries, however, the designer can apply other tolerances, such as location or orientation, and the *worst-case boundary* will always be known. This worst-case boundary is referred to as the *virtual condition*. A virtual (worst-case) condition exists whenever a size feature (such as a hole) is controlled with a geometric tolerance (such as position or perpendicularity). The virtual condition is the combined effect of size tolerance and geometric tolerances. Specific virtual conditions make holes effectively smaller and shafts effectively larger. For example, a hole has a size tolerance of between .500″ and .510″ and may be out of position as much as .005″ cylindrical zone. The virtual condition of the hole, then, is the MMC (smallest size allowed) of the hole *minus* the .005″ positional error (or .495″ diameter).

Bonus Tolerances Throughout this book, there are specific geometric tolerancing symbols and modifiers that allow *bonus (or additional) tolerances,* depending on the actual size of controlled size features. In every case where MMC and LMC modifiers have been used, bonus tolerances are allowed.

There are two terms to consider when discussing bonus tolerances: *bonus tolerance* and *additional tolerance.* A *bonus* tolerance is available on the controlled feature when MMC or LMC modifiers are used to modify the stated tolerance on that feature. MMC modifiers mean that the stated tolerance value applies only if the controlled size feature is at MMC size. If the size feature departs from its MMC size, a bonus tolerance is allowed that is equal to the amount of departure. LMC modifiers mean that the stated tolerance value applies only when the controlled size feature is at LMC size. If the size feature departs from LMC size toward MMC size, the bonus tolerance allowed is equal to that amount of departure. For example, a bonus tolerance is allowed on the position of a hole called out at MMC if the actual hole size is larger than its MMC size.

Conversely, a bonus tolerance is allowed on the position of a hole called out at LMC if the actual size of the hole is smaller than its LMC size. It is understood, in both cases, that the producer cannot violate size boundaries to obtain bonus tolerance.

Additional tolerance is a special kind of bonus tolerance that is achieved only from size datums that are called out at MMC as the actual size of the datum feature departs from its MMC size. Several examples of bonus and additional tolerances are covered in the following chapters, which include bonus at MMC, bonus at LMC, and additional tolerance from size datums.

GENERAL RULES FOR GEOMETRIC TOLERANCING

Five general rules should be understood to help the observer inspect geometric tolerances.*

Rule 1: Where only a tolerance of size is specified, the limits of size of an individual feature prescribe the extent to which variations in its geometric form, as well as size, are allowed.

The actual size of an individual feature at any cross section shall be within the specified tolerance of size.

(a) The surface or surfaces of a feature shall not extend beyond a boundary (envelope) of perfect form at MMC. This boundary is the true geometric form represented by the drawing. No variation in form is permitted if the feature is produced at its MMC limit of size.

(b) Where the actual size of a feature has departed from MMC toward LMC, a variation in form is allowed equal to the amount of such departure.

(c) There is no requirement for a boundary of perfect form at LMC. Thus, a feature produced at its LMC limit of size is permitted to vary from true form to the maximum variation allowed by the boundary of perfect form at MMC. The control of geometric form prescribed by limits of size does not apply to the following: (1) stock such as bars, sheets, tubing, structural shapes, and other items produced to established industry or government standards that prescribe limits for straightness, flatness, and other geometric characteristics. Unless geometric tolerances are specified on a drawing of a part made from these items, standards for these items govern the surfaces that remain in the "as-furnished" condition on the finished part; (2) parts subject to free-state variation in the unrestrained condition.

Perfect Form at MMC Not Required

Where it is desired to permit a surface or surfaces of a feature to exceed the boundary of perfect form at MMC, a note such as PERFECT FORM AT MMC NOT REQD is specified, exempting the pertinent size dimension from the provision of rule 1.

---

*The general rules are quoted from the standard.

8 Chapter 1 Introduction

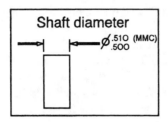

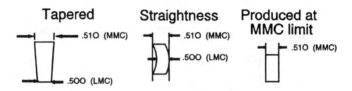

Figure 1.4 Variations in form of shafts allowed by rule 1.

Relationship Between Individual Features

The limits of size do not control the orientation or location relationship between individual features. Features shown perpendicular, coaxial, or symmetrical to each other must be controlled for location or orientation to avoid incomplete drawing requirements.

Rule 1 states that size tolerances control form (for example, roundness, straightness, and taper) and that a maximum boundary of perfect form at MMC is required. An example of a maximum boundary of perfect form at MMC is an outside diameter that must be within .500″ to .510″ (MMC is the .510″ limit in this case). If the part is produced at .510″ diameter (MMC), it must have perfect form (for example, not out of round or tapered). When inspecting size tolerances, the boundary of perfect form for a shaft outside diameter could be simulated by a ring gage at the MMC size of the shaft. Figures 1.4 and 1.5 illustrate form errors (for shafts and holes, respectively) that are allowed by rule 1.

Another case for which the form control required by rule 1 does not apply is when a size feature has a specific form tolerance. An example of this is a shaft that has a size tolerance of .500″ to .510″ *and* stated circularity, straightness, or cylindricity tolerance that is finer than the boundaries of the size tolerance (for example, circularity required within .002″).

Rule 2: For all applicable geometric tolerances, RFS applies, with respect to the individual tolerance or datum reference, or both, where no modifier has been specified. MMC or LMC modifiers must be specified where they apply.

Note: Over the years, this rule has changed significantly. In the 1973 standard, MMC was understood with position tolerance, and RFS applied to all other symbols. In the 1982 standard, MMC, LMC, or RFS had to be specified with the position symbol, and RFS was understood on all other symbols.

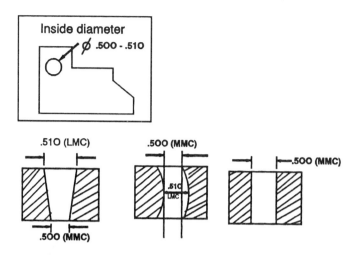

Figure 1.5 Variations in form of holes allowed by rule 1.

Pitch Diameter Rule (rule 4 in previous revisions of the standard): Each tolerance of orientation or position and datum reference specified for a screw thread applies to the axis of the thread derived from the pitch cylinder. Where an exception to this practice is necessary, the specific feature of the screw thread (such as MINOR DIA or MAJOR DIA) shall be stated beneath the feature control frame or beneath the datum feature symbol, as applicable.

Gears and Splines: Each tolerance of orientation or location and datum reference specified for gears and splines must designate the specific feature of the gear or spline to which it applies (such as MAJOR DIA, PITCH DIA, or MINOR DIA). This information is stated beneath the feature control frame or beneath the datum feature symbol, as applicable.

The pitch diameter rule applies to geometric tolerances when they are used to control features with screw threads or when screw threads are used as datum features. It states that the tolerance automatically applies to the axis of the pitch cylinder unless otherwise stated under the feature control frame or the datum symbol. In these cases, the correct feature shall be stated under the feature control frame or datum symbol.

Datum Features at Virtual Condition Rule: Depending on whether it is used as a primary, secondary, or tertiary datum, a virtual condition exists for a datum feature of size where its axis or centerplane is controlled by a geometric tolerance. In such a case, the datum feature applies at its virtual condition even though it is referenced in a feature control frame at MMC. Where it is not intended for the virtual condition to apply, an appropriate tolerance should be specified to control the form or orientation of the datum feature at its MMC.

The datum virtual condition rule applies in any case where a specific datum feature of size is being controlled with a geometric tolerance. For example, a datum hole is used to provide a datum axis for the location of three slots. However, the datum hole itself is controlled by perpendicularity tolerance to another surface on the part. The datum hole, then, has a virtual condition due to the perpendicularity tolerance that has been applied. This rule states that the virtual boundary is the basis for the location of the slots, not the MMC boundary.

FREE-STATE VARIATION

For nonrigid parts such as weldments, sheet-metal parts, thin-wall tubes, thin die castings, and many other examples, free-state variation may be applied. Free-state variation is defined as the amount a part distorts after forces that have been applied during manufacturing have been removed. For example, a thin-wall die casting is located under pressure in the jaws of a lathe and a diameter is machined in the casting. Once the machining has been completed and the casting is released, the cast material distorts (or returns to its original state before machining).

During the process, under pressure, the turned diameter may have been true (round), but after being released from the jaws, it is possible that the machined diameter will become out of round (considerably so in some cases). Free-state variation allows for this distortion in the free-state (unrestrained) condition and often calls for a roundness tolerance that may exceed the size tolerance boundaries. Whenever free-state variation is identified on the drawing, rule 1 does not apply nor does the boundary of perfect form at MMC. Size limits and form tolerances are different.

Figure 1.6 shows an example of a free-state variation application. The diameter is specified in terms of average limits. The observer should take at least four diameter measurements (90° apart) and find the average of these measurements. If the average of the four measurements falls within the average limits, the part is acceptable for size.

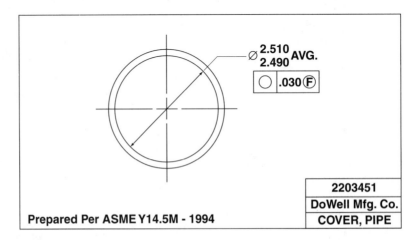

Figure 1.6 Example of a free-state variation requirement.

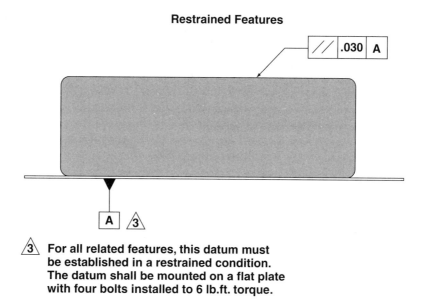

Figure 1.7 Restrained feature requirement.

The observer can also use a peripheral tape (pi tape) to measure the diameter, but the circumference measurement must be converted into its diameter equivalent, and it must be within the limits specified. As shown in Figure 1.6, if a geometric tolerance (such as roundness) is applied to a free-state part, it can and often does exceed the size tolerance. In these cases, the words FREE STATE are indicated under the feature control frame. The geometric tolerance must be inspected in the free state and must be within the stated amount of tolerance.

RESTRAINED FEATURES

In some design cases, a nonrigid part may have to be inspected in a restrained condition that simulates the part in assembly or the interface with its mating part. In these cases, the drawing will have a note that specifies the restraining condition. (An example of a restraining note is shown in Figure 1.7.) The note itself is usually flagged under the feature control frame. In this case, the part must be fixtured and torqued as indicated by the note prior to inspection.

MEASUREMENT ACCURACY AND PRECISION

Accuracy has been defined as the difference between the average of several measurements made on the part and the actual size of the part. Precision is the ability to obtain repeatable results during measurement. Both accuracy and precision are affected by measurement errors. This book will focus on measurement errors related to the types of setups covered and the equipment being used for the measurement. It is important that appreciable measurement errors be avoided by following some simple rules of measurement and geometric tolerancing rules.

There are two surfaces in a dimensional measurement, the *reference surface* and the *measured surface.* Reference surfaces on the drawings are very clear. They are called *datums.* Reference and measured surfaces are on both parts and measuring instruments. For example, a vernier (or dial) caliper has a fixed jaw and a movable jaw. The fixed jaw is the reference surface of the tool, and the movable jaw is the measuring surface. The micrometer anvil is fixed (reference surface), and the spindle is movable (measured surface).

Another important factor in measuring is discrimination. A rule of thumb used throughout the world for discrimination is the *10% rule.* This rule states that we should always make sure that the measuring instrument's discrimination consumes no more than 10% of the total tolerance of the dimension being measured. For example, a part that has a total tolerance of .001" (or ±.0005") should be measured with an instrument that discriminates to at most .0001". Another example is that a part that has a dimension with a ±.005" should be measured with an instrument that discriminates to .001", and so on. It is important to note that, if 10% of the total tolerance is a discrimination that does not exist, we use the next finer discrimination. Severe losses of accuracy and precision can result from poor selection of instrument discrimination.

Another cardinal rule in geometric tolerancing is the fact that datums must be contacted properly at all times. The earlier discussion of datums will help the observer to properly contact them during inspection. Datums and geometric rules help all observers to guarantee that their measurements are accurate, repeatable, and, most importantly, functional.

REVIEW QUESTIONS

1. A person who measures something is often referred to as the _____
2. A specific kind of dimension that is often used in geometric tolerances and has no tolerance is a _____ dimension.
3. All geometric tolerances are evaluated based on their specific tolerance _____
4. What is the MMC of a shaft that has a size tolerance of between .500" and .510"?
5. Flatness is a member of the orientation tolerance family. *True* or *False?*
6. Which general rule covers the maximum boundary of perfect form at MMC?
7. Draw a feature control frame that replaces the following note: "This surface must be parallel within .005" FIM with respect to reference surface A."
8. Draw the modifier for LMC.
9. In cases where MMC has been specified (to modify a tolerance zone), more tolerance can be allowed depending on the size of the feature controlled. This extra amount of tolerance is called _____ tolerance.
10. What is the virtual condition of a hole that has a size tolerance of between .500" and .505", and is allowed to be out of position by .005" diameter zone?
11. When a feature must be within its geometric tolerance, no matter what size, which modifier is either specified or understood?

2

Datums

OBJECTIVES

Upon completion of this chapter, the reader should be able to:

1. Have a solid understanding of datums and datum precedence.
2. Describe the difference between functional and nonfunctional datums.
3. Properly contact various types of datums for inspection.

KEY FACTS ABOUT DATUMS

1. Datums are imaginary; datum features are real part features that are used to establish datums.
2. The precedence of datums (primary, secondary, and tertiary) depends only on the left-to-right placement in the feature control frame, not on the alphabetical letter used.
3. Functional (relationship) datums are selected based strictly on the part and how it functions in assembly (not how the part is manufactured). Functional datum surfaces are often mating part interfaces in assembly or surfaces that must be related geometrically for assembly purposes.
4. Nonfunctional (construction) datums are often selected based on how the part is manufactured.
5. Simulated datums are various inspection-grade surface plate accessories (such as surface plates, gage blocks, precision parallels, gage pins, and gage rings) that are brought in contact with a part datum feature (real) to establish a datum point, line, or plane (imaginary).

6. Datum features are used "as is." For example, a surface that is out of flatness and called out as a primary datum will be contacted on the three highest points regardless of the out of flatness of the surface. At times designers will *qualify* datums by specifying an appropriate form tolerance (roundness, flatness, or the like) to improve the datum for use.

FUNCTIONAL AND NONFUNCTIONAL DATUMS

Many geometric tolerancing symbols use datum references. A datum is a point, axis, or plane derived from the true geometric counterpart of a specified datum feature. A strong knowledge of datums and how they are contacted during inspection is a prerequisite to be able to measure or gage these tolerances. A *datum* is an imaginary point, line, or plane where measurements begin. Datums are imaginary, but they are established by real *datum features* on the part. For example, a datum plane (imaginary) is established by properly contacting a specific surface on the part. A datum axis (imaginary) can be established by properly contacting a diameter on the part. Datum features are usually selected by design engineers because they define a functional relationship to another feature on the part, or they have a functional relationship in the assembly (such as interfaces or mating surfaces), or for many other reasons related to part function. Therefore, a measurement with respect to a specified design datum is a functional measurement.

There are three categories of datums: *primary, secondary,* and *tertiary.* These categories, when used, depend on the functional relationship and/or assembly requirements. In almost all cases, the primary (or first) datum *supports* the part, the secondary (second) datum *aligns* the part, and the tertiary (third) datum *stops* further movement. There is not always a need for three datums to be specified. Some applications use just one datum reference, others use two, and yet others use all three. The amount and precedence of datums called out depends on the function and/or relationship of part features in assembly. When three datums are specified, a *datum reference frame* is established that is three mutually perpendicular planes. Figure 2.1 represents the three imaginary planes for a noncylindrical part, and Figure 2.2 represents the three planes for a cylindrical part. For a cylindrical part, the axis is defined by two intersecting planes 90° apart, and an end surface is defined by one plane. Contacting cylindrical part datums depends on which is primary and which is secondary (the axis or the end surface). If the axis is primary, the two intersecting planes are established by at least three points of contact on the diameter, and the secondary plane is established by one high point of contact on the end surface. If the end surface is primary, the datum is established by at least three high points of contact on the end surface, and the secondary datum is established by one high point of contact on the diameter. Refer to Figure 2.3 for examples of contacting these datums. In cases of cylindrical parts that have a tertiary datum requirement, the tertiary datum feature is one that will stop rotation. One can always identify datums and their precedence by looking at the *feature control frame.*

The feature control frame, from left to right, contains all the necessary information. The first box contains the geometric symbol. The next box shows the geometric

Functional and Nonfunctional Datums 15

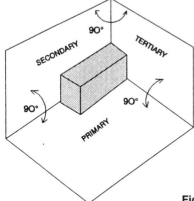

Figure 2.1 Datum reference frame for a noncylindrical part.

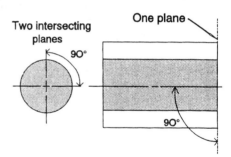

Figure 2.2 Datum reference frame for a cylindrical part.

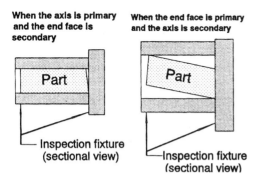

Figure 2.3 Contacting primary and secondary datums on cylindrical parts.

tolerance value (and any applicable modifiers). The next box is the place for a primary datum and modifier (if applicable). The next box is the place for the secondary datum (and modifier if applicable), and the last box contains the tertiary datum (and modifier if applicable). Examples of feature control frames are shown in Figure 2.4.

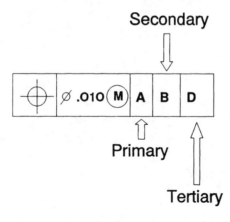

Figure 2.4 The primary datum is shown in the first datum box, the secondary datum is shown in the second datum box, and the tertiary datum is shown in the third datum box. Alphabetical order is not required.

In Figure 2.5, the example shows only one datum box at the end. This means that only a primary datum is used for this relationship. Figure 2.6 shows that the part must be located on the primary and secondary datum for the measurement, and Figure 2.7 shows that all three datums must be established for measurement. In all cases (regardless of the letter used), the primary datum is in the first box, the secondary datum is in the second box, and the tertiary datum is in the third box.

The feature control frame establishes the specific geometric relationship to be measured. An example of a complete feature control frame is shown in Figure 2.8. The first box always contains the geometric symbol of control. The next box to the right contains the geometric tolerance (including any modifiers on the tolerance), and the datum boxes are last. The feature control frame describes the relationship without need for long notes on the drawing.

Figure 2.5 Feature control frame with a primary datum.

Figure 2.6 Feature control frame with primary and secondary datums.

Figure 2.7 Feature control frame showing all three datums.

SIMULATED DATUMS

The latest revision of ASME Y14.5M (1994) covers the topic of simulated datums. Simulated datums are inspection equipment that can be used to contact part datum features and establish datum points, lines, or planes. The best example of a simulated da-

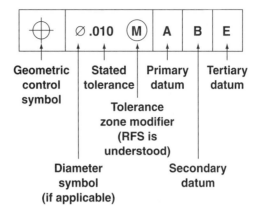

Figure 2.8 Feature control frame for a position tolerance.

tum (for planes) is the surface plate. A part that has a datum surface can be brought into contact with the surface plate and establish a datum plane. Another example is a part that has an outside diameter as a datum feature. The outside diameter can be placed in a vee block (simulated datum) or in a collet to establish a datum axis. Another example is a slot that is identified as a datum. The centerplane (or datum) of the slot can be established using a gage block stack that fits the slot.

A final example is a part that has a hole as a datum feature. Datum holes are best contacted with the largest gage pin that can be inserted into the hole. In this case, the datum is an imaginary axis that is established using a datum feature (the hole) and a simulated datum feature (the pin). Other examples of simulated datums are shown in various chapters in this book, such as right-angle plates (knees), sine plate surfaces, gage block stacks, gage pins, and many other examples of typical inspection equipment. All this equipment can be used as simulated datums to contact part datum features for inspection.

QUALIFIED DATUMS

In some cases, a datum feature (or datum feature of size) must be *qualified* in order to provide the proper reference for the geometric control being applied. An example of a qualified datum is where a surface has been identified as a datum for a parallelism control and above the datum symbol is a flatness requirement (refer to Chapter 4). This indicates that the datum feature must be evaluated for flatness *prior* to the parallelism measurement. In this case, flatness *qualifies* the datum for use. Other examples of qualified datums are (1) a datum diameter with a cylindricity requirement, (2) a noncylindrical datum feature of size that has a parallelism requirement, and (3) a datum hole with a roundness requirement. In any case, when inspecting geometric tolerances that use datum references, the observer should make sure that the datum qualifiers are inspected prior to contacting the datum for other measurements. It is equally important to understand that datums that are *not* qualified by specific controls on the drawing are contacted *as is* during inspection.

CONTACTING FUNCTIONAL DATUMS

There are two kinds of datums, functional and nonfunctional. Functional datums are those datum features called out on an engineering drawing that are directly related to part fit, form, and function in the assembly. Nonfunctional datums (for example, datum targets) are datum features that usually have little to do with part function. Nonfunctional datums, for example, are the datum targets used on a casting for the purpose of identifying how the casting shall be contacted at machining operations.

Functional datums are identified on an engineering drawing in a box with any letter of the alphabet except I, O, or Q. Figure 2.9 shows an example of a functional datum symbol. Functional datums are always contacted at their highest points (preferably a simulation of the kind of contact the part makes in assembly). Nonfunctional datums are always contacted at specified points identified by datum target symbols.

Figure 2.9 Functional datum symbol.

Primary functional datum features must be contacted, at minimum, on the three highest points. A primary (support) datum on a surface is established by bringing the datum surface into contact with a simulated datum plane (a surface plate, for example). A primary datum axis (such as the axis of a hole) is best established by the axis of the largest gage pin that can be fully inserted into the hole. In this manner, the actual contact that is made will be at least the three (or more) extreme points inside the hole. Secondary functional datum features (used to align the part) must be contacted at the two highest points. A secondary (aligning) datum plane, for example, is contacted by bringing the secondary datum surface in contact with a true surface (such as a right-angle plate, or *knee* as it is called in the shop).

The secondary datum plane is perpendicular to the primary datum plane; therefore, the knee can be used on a surface plate to establish primary datums (on the plate) and secondary datums (on the knee). Tertiary (stopping) datums are established by contacting a minimum of one point on the datum feature.

A tertiary datum plane can be contacted by bringing the tertiary datum feature in contact with a true plane (such as a knee or a precision parallel). The tertiary datum plane is perpendicular to both primary and secondary datum planes; therefore, the simulated datum used must be perpendicular to the other simulated datums.

On a noncylindrical part, once the primary, secondary, and tertiary datums are established (if all three are called out on a drawing), the part is ready for inspection of the tolerances that have been called out to those datums. With all three datums contacted, movement of the part is restricted such that inspection can begin, and the inspection will be functional if they are functional datums. With this kind of setup, the three imaginary datum planes are called the *datum reference frame*. Keep in mind that, depending on the function of the part and the application, only one or two datums may be called out on the drawing. Not all functional relationships require all three datums. In fact, many applications use just the primary datum.

With respect to cylindrical parts, the function of a datum is the same as previously discussed (support, align, and stop); only the description of the function is slightly different. On a cylindrical part, the primary datum supports the part, the secondary datum aligns the part, and the tertiary datum *stops rotation*. Also, the primary datum may be the axis of the diameter, and the secondary datum may be an end surface of the diameter (or vice versa). In all cases, it is important to understand that the datums are contacted in order of precedence (primary first, then secondary, then tertiary).

COMPOUND DATUMS

Compound datums are planes, axes, points, or lines that are established using more than one datum feature on a part. For example, a datum axis could be established by contacting two different diameters (outside or inside diameters) on the same part. A datum plane could be established by contacting two different surfaces on the part. In these cases, the feature control frame will identify compound datums, as shown in the examples in Figure 2.10

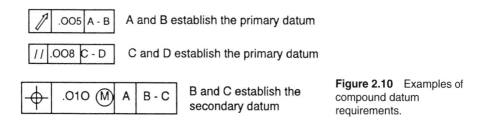

Figure 2.10 Examples of compound datum requirements.

Compound datums are easily recognized by the use of two different datum letters separated by a dash and placed in a single datum box at the end of the frame. If the plane or axis being established is a primary datum, these two letters will be in the first datum box; if it is a secondary datum, the two letters will be in the second box; and so on. Figure 2.11 shows a part drawing with an example of a primary compound datum.

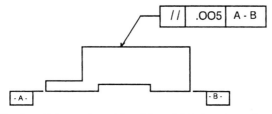

The three highest points of A and B establish the primary datum

Figure 2.11 Drawing with a primary compound datum.

OFFSET COMPOUND DATUMS

In some cases, two offset surfaces on a part may be involved with the function of the part to the extent that they are used to establish a datum plane. In such a case, the surfaces will be identified with different datum letters, and the feature control frame will call out these letters with a dash in between (as previously discussed). But there will be one more dimension on the drawing.

A basic dimension must be used to identify the exact offset of two surfaces that are not in the same plane when they are used as compound datum features. The drawing in Figure 2.12 shows an example of compound offset datum features. The .650" basic dimension defines the offset required between the two simulated datums that will be used to establish the datum plane.

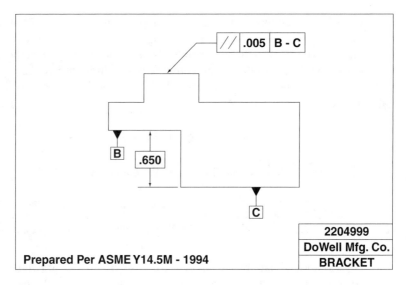

Figure 2.12 Example of compound offset datum features.

EQUALIZING DATUMS

On products that have contoured or rounded edges, for example, observers may have to contact datums that *equalize* the part. Figure 2.13 shows an example of a part for which the datum plane must be contacted using equalizing datums. A set of aligned vee locators that are each 60° can be used to establish the datum plane in this case. Datum targets are used to identify the type of contact that must be made and the basic angle of the vee locators.

Datum Targets (Nonfunctional Datums)   21

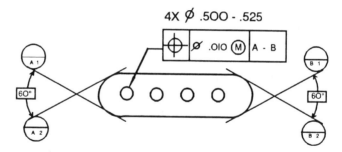

Figure 2.13 Equalizing datum that would require matched vee locators.

DATUM TARGETS (NONFUNCTIONAL DATUMS)

Datum targets are usually nonfunctional datums (that is, contacting them has no relationship to how the finished part works in assembly). Datum targets are mostly used to establish datums for machining castings, bar stock, forgings, weldments, and other materials. Their primary function is to establish *consistency* in the way these materials are located in machining operations and to maintain consistency in how these materials are inspected in the as-cast or as-forged condition. The one exception when targets are functional is when datum target symbols are used on the drawing in conjunction with a functional datum symbol.

There are three kinds of datum targets: target points, lines, and areas. A datum target point is contacted at the specified location with contact pins that are pointed or have a radius at the point of contact. A datum target line is contacted typically with a knife edge or the outer edge of a round pin. A datum target area is usually contacted by the end of a pin that has the specified diameter (or square) area. Figure 2.14 shows datum target symbols for target points, lines, and areas.

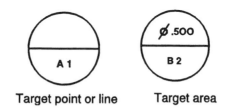

Figure 2.14 Datum target symbols.

The use of primary, secondary, and tertiary datums still applies to datum targets except that it is no longer the highest points that must be contacted. A primary (support) datum target must be established by contacting the *three specified points* identified and dimensioned on the drawing. A secondary (alignment) datum target must be established by contacting the *two specified* points identified and dimensioned on the drawing. A tertiary (stopping) datum target must be established by contacting the *one specified* point identified and dimensioned on the drawing.

The primary, secondary, and tertiary datum targets will be identified by the datum target symbols using different letters. Again, the letters I, O, or Q are not used. For example, primary datum targets may use letters A1, A2, and A3 (for three points of contact), and secondary datum targets may use letters C1 and C2 (for two points of contact), and the tertiary datum target may be B1 (for one point of contact).

In some cases, the primary datum is target points, the secondary datum is target areas, and the tertiary datum is a target line (or other combinations are used). Regardless of the targets called out on the drawing, it is up to the observer to contact them properly before the inspection (or layout) takes place. An example of datum targets is the box casting shown in Figure 2.15. In this case, datum targets have been called out for locating the box casting in the first machining operation. Each target is identified by the datum target symbol (for example, point, line, or area). Also, the targets would be dimensioned in order to tell the observer their location and the tolerance on the location (not shown in Figure 2.15). If targets are dimensioned with simple plus or minus tolerances, these tolerances would be used.

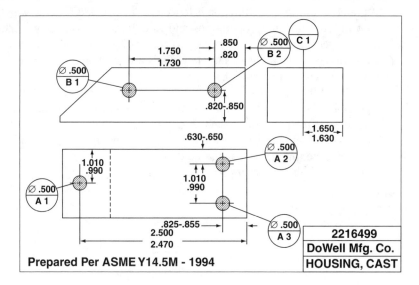

Figure 2.15 Datum targets called out on a casting.

If targets are dimensioned with *basic* dimensions, it is understood (per the standard) that tooling tolerances apply to the location of the targets for machining and inspection purposes.

Figure 2.16 shows an example of how the box casting would be targeted for scribing layout lines in inspection. Not shown are the clamps that should be used to hold the casting against the target pins for layout. This is also an example of the locating pins that would be used at the first machining operation for the casting.

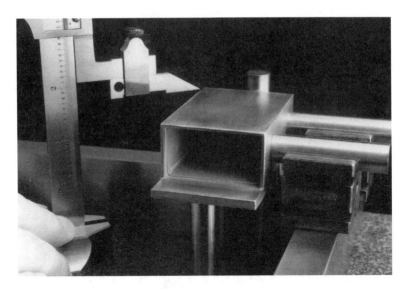

Figure 2.16 Setup for inspecting the box casting to datum targets.

In summary, most geometric tolerances use specific datums because they require a datum reference and because the datums are related to fit, form, function, or performance of the part in assembly (functional datums) or they are related to obtaining consistent results in machining operations (nonfunctional datums).

REVIEW QUESTIONS

1. Points, lines, axes, or planes that are used as a reference for measurement are called _____.
2. The purpose of a primary datum is support. *True* or *False?*
3. If a datum surface has not been qualified with a form tolerance, the observer should contact the datum surface as is. *True* or *False?*
4. Datums are imaginary, but datum features are real. Datum features are used to establish datums. *True* or *False?*
5. A primary datum surface must be contacted at three or more highest points on the surface. *True* or *False?*
6. Datum _____ are usually nonfunctional and are contacted at specified points.
7. Provide a sketch that shows how a drawing would specify compound datums A-B in a feature control frame.
8. A functional tertiary datum surface must be contacted at the _____ point on the surface.
9. A secondary datum target surface must be contacted at two _____ points.

3

Inspecting Size Tolerances

OBJECTIVES

Upon completion of this chapter, the reader should be able to:

1. Understand that size tolerances, per the ASME standard, control more than just size.
2. Understand the maximum boundary of perfect form at MMC.
3. Properly measure (or gage) size tolerances.
4. Understand boundary gaging and be able to simulate boundary gages using surface plate accessories.
5. Understand the limitations of hard gages (such as plug gages and ring gages).

INTRODUCTION

Size tolerances (such as outside and inside diameters, thickness, and length) have often been inspected using either go/no-go gages or standard measuring instruments such as calipers and micrometers. In many cases, due to the precision of the process, simple measurements of size tolerances did not cause appreciable error in the acceptance or rejection decisions that were made because the boundaries of size (MMC and LMC) were not violated or because the part was so close to nominal size that it was not a borderline decision.

It is important for the observer to understand that size tolerances are governed by the provisions of rule 1 (previously discussed in Chapter 1). To properly (and functionally) inspect size tolerances both boundaries should be inspected.

Another problem is that, in most companies, either the functional hard go/no-go gage is used *or* the measuring instrument. The problem lies in the fact that, without exploration, most measuring instruments (such as micrometers and calipers) cannot verify the maximum boundary of perfect form at MMC required by rule 1, and, conversely, most go/no-go gages cannot verify the LMC boundary at any cross section. This presents a dilemma in the use of measuring instruments or hard gages. The problem can be overcome by using both types of gaging in those cases where rule 1 applies. Those who use measuring instruments will not necessarily have to purchase go/no-go gages. Functional measurements can be made using surface-plates and surface-plate accessories found in most machine shops or inspection areas. Some examples of functional measurements are covered in the following paragraphs.

For functional inspection of size tolerances to take place, we must understand the *inherent control* required by rule 1 of the standard. For example, a simple thickness tolerance on a noncylindrical part (per rule 1) automatically controls the *parallelism* of the opposing faces involved. Size tolerances on noncylindrical part features inherently control form errors such as straightness, bell-mouthed conditions, hourglass conditions, and taper. Few standard measuring instruments (such as micrometers, calipers, indicators, or probes) can verify both the LMC and MMC conditions of size features. These instruments can usually be used to inspect only one, not the other.

NONCYLINDRICAL PARTS

For noncylindrical parts, size dimensions are generated by two surfaces (the reference and measured surface). Therefore, the MMC boundary of perfect form is defined by two imaginary planes at MMC distance apart. These two imaginary planes could be simulated by two precise boundaries at MMC distance apart, such as two precision parallels for outside thickness measurements or an adjustable parallel for inside measurements (such as a slot). For example, refer to the part in Figure 3.1. This part has an overall width tolerance of .188 to .192. Rule 1 states that the part cannot violate its maximum boundary of perfect form at MMC. In this case, the company that uses go/no-go gages only will *not* be able to verify undersize conditions because gages cannot see these conditions at each cross section. Also, the company that uses micrometers will not be able to verify the maximum boundary of perfect form because the micrometer is not equipped to simulate two parallel planes the MMC distance apart. The following equipment (or other similar equipment) can be used to inspect the part for both conditions.

Inspection of the maximum boundary of perfect form at MMC for the part can be accomplished with a go/no-go gage as long as the gaging faces completely surround both surfaces of the dimension and they are at the part's MMC size. If there is no gage, there are other ways of establishing a simulated MMC boundary with standard surface plate accessories.

One example is shown in Figure 3.2. In this case, two precision parallels on a surface plate have been separated by two gage block stacks that are the same as the MMC boundary of the part. If the part dimension will pass between the two parallels without

26 Chapter 3 Inspecting Size Tolerances

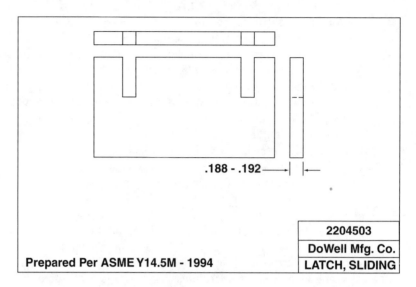

Figure 3.1 Size tolerance on a part drawing.

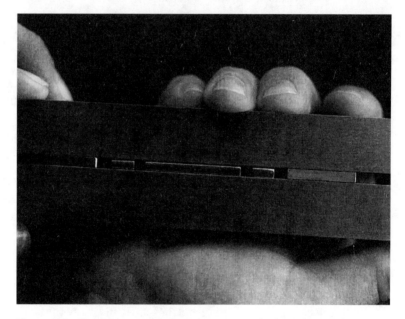

Figure 3.2 Precision parallels have been set up for boundary gaging.

resistance, the boundary of perfect form has not been violated. Then a micrometer could be used at sections of the part dimension to inspect for undersize conditions.

Another procedure for inspecting the maximum boundary of perfect form (using variables) is the surface plate. A different part (height dimension) is used for this example. The surface plate, an indicator (mounted to a surface gage or height gage), and a gage block stack can be used to verify the MMC boundary. First, select an indicator that discriminates to no more than 10% of the total tolerance of the dimension. Next, wring a gage block stack to the MMC size limit of the part. Then set the gage block stack, the part, and indicator on a surface plate and zero the indicator on the stack as shown in Figure 3.3. Now zero represents the MMC limit of the part. Next, pass the indicator over the entire surface of the part as shown in Figure 3.4 and look for any readings higher than zero. If sweeping the part does not show plus readings above zero, the part is within its MMC boundary of perfect form. Once the maximum boundary of perfect form has been verified, a micrometer can be used to measure cross sections of the part for undersize conditions, as shown in Figure 3.5.

Figure 3.3 Set zero on the gage block stack.

Figure 3.4 Sweep the entire part surface looking for readings higher than zero.

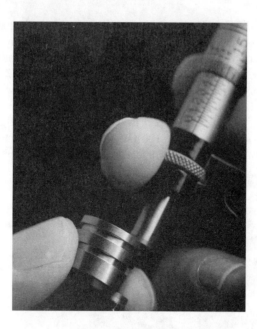

Figure 3.5 Checking for undersize conditions with a micrometer.

CYLINDRICAL FEATURES: SHAFTS

For cylindrical features, the MMC boundary of perfect form is an imaginary cylinder at MMC size. This cylinder can be simulated by ring gages (for shafts) and plug gages (for holes) since these gages are true cylinders within gage tolerances. Therefore, plug gages verify the MMC boundary of perfect form for holes, and ring gages verify the MMC boundary of perfect form for shafts.

Consider the shaft drawing in Figure 3.6. The shaft has an MMC and LMC boundary of size per its allowable size limits. The largest size limit is MMC, and the smallest size limit is LMC. The maximum boundary of perfect form, then, is MMC size. This boundary can easily be verified by using a ring gage that has an inside diameter equal to the MMC diameter of the shaft. The ring gage should also encompass the entire shaft at one time or the MMC boundary will not be verified. Figure 3.7 shows the shaft's MMC boundary being inspected with a ring gage.

Once the MMC boundary of perfect form for the shaft has been inspected with the ring gage, the shaft diameter elements can be measured with a micrometer in various places to ensure that the shaft is not undersize at any cross section. Figure 3.8 shows the observer exploring the shaft for undersize conditions with a micrometer.

CYLINDRICAL FEATURES: HOLES

The same measurement techniques apply for holes, except in reverse. For a hole, the boundary of perfect form is the smallest hole size allowed. One gage that is commonly used for MMC boundaries of holes is a typical plug gage. For example, the drawing in Figure 3.9 shows a hole and its size tolerance. The smallest size is the MMC boundary.

Cylindrical Features: Holes   29

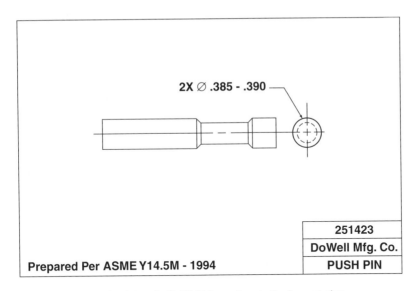

Figure 3.6 Drawing for a shaft; MMC boundary is the largest size.

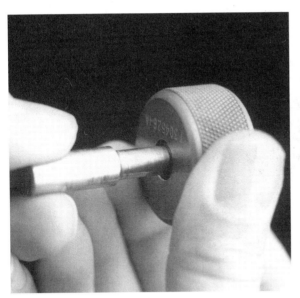

Figure 3.7 Ring gage checks the MMC boundary of the shaft.

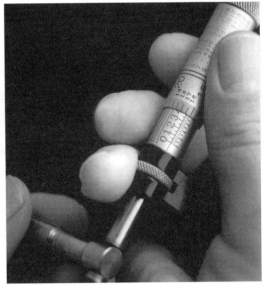

Figure 3.8 Micrometer checks the LMC boundary of the shaft.

30 Chapter 3 Inspecting Size Tolerances

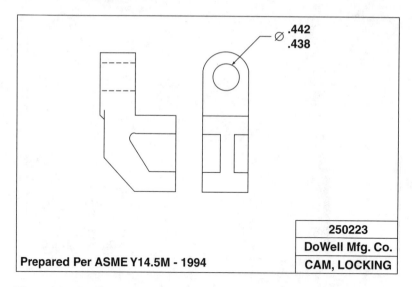

Figure 3.9 Drawing for a hole; MMC boundary is the smallest size.

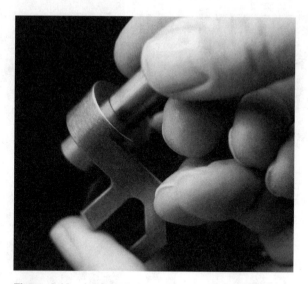

Figure 3.10 A gage pin is used to check the MMC boundary of a hole.

To inspect the boundary of perfect form at MMC for the hole, a plug gage (or a gage pin) can be used as long as the gage will extend through the entire length of the hole and it is made at the MMC size of the hole (see Figure 3.10). If the plug gage will go into the hole without resistance, the MMC boundary of perfect form is acceptable.

Cylindrical Features: Holes

(a)

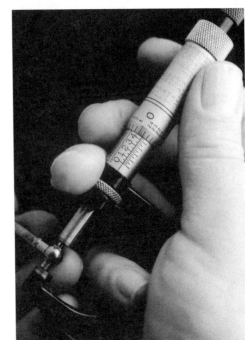

(b)

Figure 3.11 (a) A telescoping gage is used to transfer the LMC boundary of a hole, and (b) a micrometer is used to measure the telescoping gage.

The plug gage (or gage pin) is limited to inspecting one boundary, MMC. Plug gages for holes (like ring gages for shafts) cannot be used to inspect the LMC boundary. In cases of holes, however, various hand-held gages can be used to look for oversize conditions, such as dial bore gages, telescoping gages, and small hole gages. Figure 3.11a and b show a telescoping gage being used to look for oversize conditions of the hole.

PROBLEMS WITH GAGES

Hard gages such as ring gages, plug gages, and functional gage fixtures are functional because they provide a physical representation of the worst-case mating part. They are not without problems, however. Some of these include wear, gage-maker tolerances, cost, maintenance, and storage. For example, these gages are always working *against* the part, and they must be designed for wear (wear allowance). Also, the gage maker must have some tolerance to make the gage (even though gage tolerances are quite small). Due to wear allowance and gage-maker tolerances, these gages will often be the *same size as the MMC boundary* that they are being used to inspect. To compound this problem, at the time when the gage has worn down enough to make the exact decision, it is often out of calibration. A cardinal rule with hard gages is *if the gage rejects the part at the borderline, always measure the part with a measuring instrument to make sure the part is, or is not, acceptable. Gages are designed to reject borderline parts!*

Another limitation of hard gages is that they cannot be used to inspect the LMC boundary of a feature of size. For example, a plug gage cannot detect certain conditions of a hole, such as taper, out of round, and crookedness, nor can a ring gage detect these conditions on a shaft. Other examples of diametral conditions are hourglass-shaped, barrel-shaped, and bell-mouthed diameters. There are also specific roundness problems on diameters such as three-, five-, and seven-lobed parts where these hard gages can measure only the *effective size,* not the actual size. (Refer to Chapter 6 for more information on lobes.)

ALTERNATIVES TO HARD GAGES

There are many ways that standard surface plate accessories (or hand tools) can be used when hard gages are not applicable, available, or affordable. Refer to the slot dimension on the drawing in Figure 3.12. In Figure 3.13a, an adjustable parallel is used in conjunction with a micrometer to inspect the MMC boundary of the slot. In Figure 3.13b, a gage block stack (stacked at the MMC boundary of the slot) is used for MMC boundary gaging.

For holes without dedicated plug gages, a set of gage pins can be used to inspect the MMC boundary; for example, the part in Figure 3.14 can be inspected with a pin as shown in Figure 3.15.

Alternatives to Hard Gages

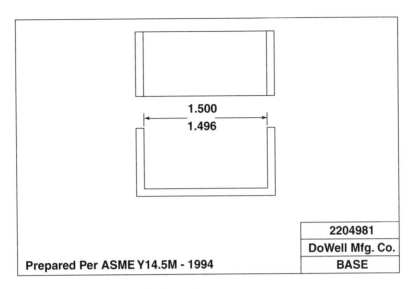

Figure 3.12 Size dimension for a slot.

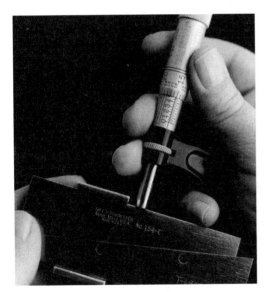

(a)

(b)

Figure 3.13 (a) An adjustable parallel and a micrometer used to measure the MMC boundary of the slot; (b) a gage block stack can also be used to gage the MMC boundary of the slot.

34 Chapter 3 Inspecting Size Tolerances

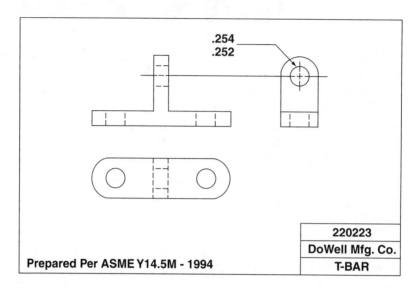

Figure 3.14 Hole size tolerance.

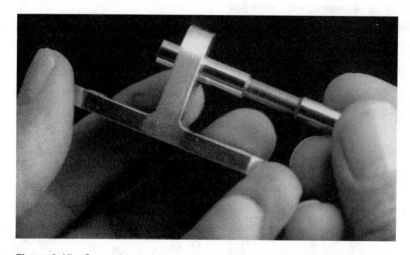

Figure 3.15 Gage pin used to check the MMC boundary of a hole.

SUMMARY

One of the most important aspects brought out by this chapter is the fact that size features need to be functionally inspected. A hole should be inspected in a way that simulates or reproduces the mating part (for example, the pins or bolts that go into the hole when assembled). A shaft, pin, or stud should be inspected in a way that simulates or reproduces the mating part (for example, the hole the shaft, pin, or stud goes into when assembled). Another important part of functional inspection is the fact that traditional gages or measuring instruments alone *do not* provide for functional representation of the mating part.

REVIEW QUESTIONS

1. Per rule 1 of the standard, simple size tolerances control _____ as well as size.
2. All size features for which rule 1 applies have a required _____ of perfect form at MMC.
3. Hard size gages (such as plug and ring gages) are typically limited to gaging the MMC boundary, but cannot gage the _____ boundary.
4. Hard gages (such as plug, ring, and functional gages) are designed to reject borderline _____ parts at times.
5. Hard gages alone cannot be used to completely inspect size tolerances when rule 1 applies. *True* or *False?*
6. Surface plate accessories (such as precision parallels and gage blocks) can often be used to verify the boundary of perfect form at MMC. *True* or *False?*
7. A shaft that has a size tolerance of between .500″ and .502″ (when rule 1 applies) must also be round within the boundaries of size. *True* or *False?*
8. Two reasons why a "go" plug gage for a hole is often the *same* size as the low size limit of the hole are (1) gage-makers' tolerances and (2) _____ allowance.
9. For a shaft size (per rule 1), the observer could use a micrometer to measure the LMC boundary of size and a _____ gage to gage the MMC boundary of size.
10. Some size features can be inspected for their MMC boundary using a surface plate, a gage block stack, and a dial indicator. *True* or *False?*

4

Flatness

OBJECTIVES

Upon completion of this chapter, the reader should be able to:

1. Interpret flatness tolerances and the flatness tolerance zone.
2. Apply various fixed-plane methods for inspecting flatness, such as Indian pins and surface plates with probes.
3. Apply various optimum-plane methods for inspecting flatness, such as jackscrews and wobble plates.
4. Use optical flats for tight tolerance flatness inspection, and interpret interference light bands for flatness error.
5. Understand that flatness error cannot violate size tolerances for associated size dimensions.

INTRODUCTION AND APPLICATIONS

Flatness is a form control that applies to a single continuous surface. Although flatness is never specified to a datum feature, it has an intrinsic datum called an *optimum plane*. It is understood that when a feature control frame indicates a flatness requirement on a surface of the part, the measurement must be made with respect to the *full indicator movement (FIM)* found on an indicator with respect to the optimum plane.

To inspect flatness, the observer must establish an optimum plane and make a FIM measurement with reference to the plane. There are many functional reasons for flatness requirements on engineering drawings. Some of these are (1) to ensure the integrity of mating (or mounting) surfaces, (2) to qualify surfaces that are datums, (3) to ensure that surfaces seal properly, and (4) to ensure that a proper seal is maintained.

Flatness cannot be gaged (or evaluated in a go/no-go manner); it must be measured. We can, however, design a gage with indicating probes that will measure flatness as long as the discrimination of the indicator is appropriate with respect to the flatness tolerance (for example, the 10% rule), and the gage designed can establish an optimum plane for measurement (such as a built-in leveling plate). Surface plate methods can also be used to measure flatness.

KEY FACTS

1. Datums references are not applicable to flatness. Flatness is to itself (intrinsic datum).
2. Flatness cannot be gaged; it must be measured.
3. Measurements of flatness must be focused on the hills and valleys of the surface and must reproduce the tolerance zone of two parallel planes the stated tolerance apart.
4. Flatness is usually applied to continuous surfaces (for interrupted surfaces, refer to Chapter 14).
5. Permissible error in the flatness of a surface cannot be added to size tolerances associated with that surface.
6. A flatness tolerance must be less than the associated size tolerance.

Example 1: Flatness Requirement

A typical drawing requirement for flatness is shown in Figure 4.1. The feature control frame states that the surface of the part must be flat within .001" (FIM is understood). The tolerance zone for the flatness requirement is defined as *two imaginary parallel planes that are .001" apart*, as shown in Figure 4.2. The part in Figure 4.1 will be used as an example in the following methods for inspecting flatness.

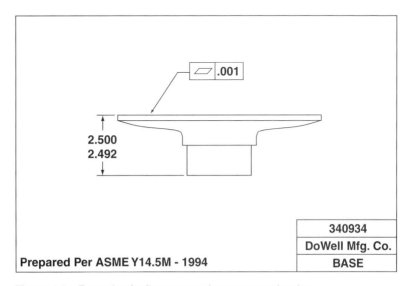

Figure 4.1 Example of a flatness requirement on a drawing.

Chapter 4 Flatness

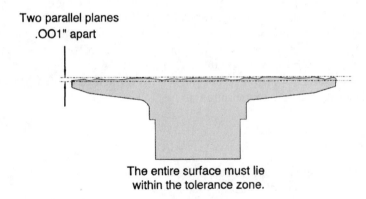

Figure 4.2 Tolerance zone for the part.

FLATNESS MEASUREMENT: JACKSCREW METHOD

One set of accessories for a surface plate is leveling (jack) screws. This method is widely used in industry to establish the optimum plane for flatness measurement. The following is a list of the equipment needed for measuring flatness using jackscrews.

1. Surface plate
2. Dial indicator with proper discrimination
3. Set of three jackscrews
4. Mount for the indicator (height or surface gage)

STEP 1 Prelevel all three jackscrews using a height gage (or another tool of constant height from the surface plate), as shown in Figure 4.3. This preliminary step is only to start the jackscrews on an equal-height basis for the setup. The jackscrews will be used to level the part surface with respect to the surface plate surface. The leveling takes less time to set up when the jacks start out at equal height.

STEP 2 Set the jackscrews on the surface plate in a triangle, and mount the part on top of the jackscrews such that the surface to be measured is exposed (see Figure 4.4).

STEP 3 If the part surface will not be damaged, use a nonpermanent marker to place a small mark on the surface to be measured (directly above each jackscrew). This step is necessary so that when leveling the part, the observer can find the same location above each jackscrew. (*Caution:* Avoid obvious flaws in the surface when marking these three spots.)

STEP 4 Beginning at any one of the marked locations, set zero on the dial indicator and move to another marked location. (*Note:* Avoid possible cosine errors by making sure the dial indicator tip is as parallel as possible to the part surface.) Then use the jackscrew to raise or lower the part and obtain a zero

Flatness Measurement: Jackscrew Method 39

Figure 4.3 Preleveling the jackscrews.

Figure 4.4 Part is mounted on preleveled jackscrews.

reading at *each* marked location. Continue moving between the three locations until zero is set at all three (as shown in Figure 4.5). (*Caution:* Avoid bumping the part during or after the setup. Otherwise, zero at each location will have to be reset. Once the dial indicator reflects zero at all three locations, the optimum plane has been established.)

40 Chapter 4 Flatness

Figure 4.5 Part surface has been leveled to the surface plate.

Figure 4.6 Comparing the surface to the optimum plane (FIM).

STEP 5 Move the dial indicator all over the controlled surface and find the resultant FIM of flatness, as shown in Figure 4.6. The full indicator movement should not exceed the flatness tolerance indicated in the feature control frame.

In summary, the observer has used the jackscrews to level the controlled surface parallel to the surface plate so that parallelism error was nullified and an optimum plane (established by the three locations at zero) was generated. Once the optimum plane was generated, the peaks and valleys of flatness error were measured by the FIM on the indicator. This method can also be used on small parts by using an intermediate plate that will cover the jackscrews, placing the small part on the plate, and then establishing zero in three areas of the part surface in line with the jackscrew locations. Keep in mind that anything between the jackscrews and the surface of the part being controlled does not add error to the measurement because these intermediate surfaces and their error are nullified by the leveling of the part surface to the surface plate.

FLATNESS MEASUREMENT: WOBBLE-PLATE METHOD

Leveling *wobble* plates, as they are often called, can be used instead of jackscrews to establish the optimum plane for the part. Similar to jackscrews, a wobble plate has three leveling jackscrews built into it and a direct mounting surface to rest the part on (see Figure 4.7).

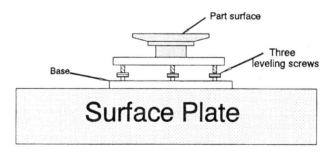

Figure 4.7 Leveling wobble plate in use.

The method for using a wobble plate is the same as the jackscrew method except, for smaller parts, the three marks made may have to be in line with the leveling screws of the plate, instead of directly above them. Once again, level the part using the leveling screws and find the FIM of the measured surface.

FLATNESS MEASUREMENT: FIXED-PLANE METHODS

Another method that can be used to establish the optimum plane for flatness measurement is what the author calls the *fixed-plane method*. In this case, the optimum plane is established by using three accessories for the surface plate that are equal in height within gaging tolerances. Examples of these are (1) three equal-height gage block stacks, (2) three equal-height pins (often called *Indian pins*), or (3) three equal-diameter precision balls. In each case, an imaginary optimum plane is established within gaging tolerances. Once this plane is established, the surface to be measured for flatness is placed directly on the accessories used.

The equipment needed for this method is as follows:

1. Surface plate
2. Dial indicator of proper discrimination
3. Height or surface gage
4. Three equal-height accessories (pins, gage block stacks, or the like)

For best results, most observers prefer three equal-height Indian pins that each have a radius at the point of contact with the part surface. The Indian pins allow for a more thorough indication of the surface during the measurement.

STEP 1   Set the three equal pins, blocks, or other accessories on the surface plate in a triangle at a distance apart that will allow the part to be measured to rest on them as shown in Figure 4.8. (*Note:* In Figure 4.8, three equal pins are used.)

Figure 4.8   Part surface to be measured rests on the optimum plane established by three equal pins.

STEP 2   Carefully place the part surface to be measured *directly* on top of the three accessories such that the observer can still access the surface to be measured with the dial indicator.

STEP 3   Invert the dial indicator so that FIM measurements can be made on the underside of the setup (as shown in Figure 4.9), and traverse the indicator over the entire controlled surface, watching for the FIM. The FIM should not exceed the tolerance indicated in the feature control frame.

Figure 4.9 FIM being obtained from the inverted surface.

In this case, the observer established the optimum plane using three accessories (pins) that were equal height within gaging tolerances. Gage blocks or precision balls of the same size could also have been used. In this manner, it was possible to compare the controlled surface to an imaginary (and perfectly flat) plane established by the accessories. The full indicator movement measured the peaks and valleys of flatness of the surface. The major limitations of the fixed-plane method are (1) greater difficulty accessing the surface (inverted) with the dial indicator, (2) surface not completely accessible due to the surface areas covered by the pins, (3) difficulty in keeping Indian pins and other accessories at equal height within gage tolerance, and (4) possibility of damage to the part since the method is in direct contact with the surface to be measured.

FLATNESS MEASUREMENT: DIRECT-CONTACT METHOD

Another method for measuring flatness is direct contact with a simulated plane (such as a surface plate) that has a built-in probe or indicator. The surface plate shown in Figure 4.10 is designed to measure flatness by placing the controlled surface directly on the surface plate.

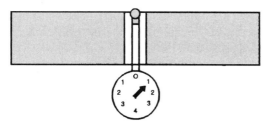

Figure 4.10 Surface plate with a built-in probe to check flatness.

Chapter 4 Flatness

The equipment needed for this method is as follows:

1. Special surface plate with a built-in probe or indicator of proper discrimination
2. Flat surface within gaging tolerances to be used to *master* the indicator (or probe)

STEP 1 The surface plate must first be mastered at zero on the probe (or indicator). This is accomplished by using a flat surface within gaging tolerances, such as a gage block or a precision parallel, as shown in Figure 4.11.

STEP 2 Once the zero setting has been made, the part should be rocked gently onto the surface plate; then, probing the entire surface of the part, watch for the full indicator movement (FIM), as shown in Figure 4.12. The FIM should not exceed the flatness tolerance shown in the feature control frame.

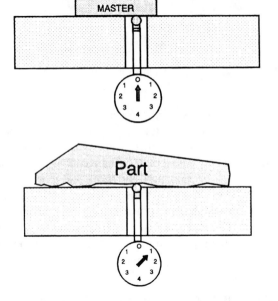

Figure 4.11 Mastering the probe (indicator) with a precision parallel.

Figure 4.12 Inspecting the part.

Although the contact methods previously mentioned are faster than leveling methods, there are some limitations.

1. The part surface to be measured is in full contact with the surface plate; therefore, high areas will raise the part from the probe or indicator and then lower the part when these areas pass over the probe. In such cases, the direct-contact method makes the part appear to be slightly worse than the true value of flatness error.
2. Using any method of direct contact offers the possibility that the part may be scored or damaged.
3. In the case of the surface plate and probe, a special surface plate must be purchased with a built-in probe. Jackscrews are less expensive and can be used on a wide variety of products.

OPTICAL FLATNESS INSPECTION

Optical flats are often used for measurement of flatness in very tight tolerances (for example, .0001" or less) on surfaces that have been lapped or polished (such as gage blocks, micrometer anvil and spindle faces, snap gage faces, seals, valve seats, and other precision flat-lapped parts). Optical flats are made from clear fused quartz or, at times, Pyrex glass. Some optical flats are coated to block out other light sources. Optical flats can be purchased with various degrees of accuracy from reference grades that are flat within about 1 millionth to commercial grades that are flat to about 4 millionths.

To use optical flats, the observer must use the correct procedure and be able to read *interference light bands* (called *fringes*) that result from bringing the flat in contact with the part. A monochromatic light source (such as helium) is also required for measurement of flatness with optical flats. Part surfaces must have a finish that will reflect light (such as fine-ground or lapped surfaces) in order for optical flats to be used.

Interference Light Bands (Fringes)

When an optical flat (hereafter referred to as the *flat*) is brought in contact (with a very small wedge) with a lapped part surface, dark bands (or *fringes*) will appear. Each band represents an interval of about 11.57 millionths of an inch from the surface being inspected to the flat (when the bands are viewed at 90° from the surface of the flat). This is especially true when a monochromatic light source is used.

Procedure for Testing Flatness with Optical Flats

STEP 1   Make sure the part surface (and the flat) are clean and the part surface is free of burrs and nicks. Remove dust from the part and the flat with a camel hair brush (or with the palm of the hand), and remove burrs or nicks from the part surface with an appropriate deburring stone.

STEP 2   Place the part surface to be inspected under a monochromatic light source and place a clean piece of paper (optical tissue can be purchased) over the surface to be inspected. The primary purpose of the paper is to protect the part surface and the flat from being scratched.

STEP 3   Place the flat on top of the paper and, holding the flat steady with one hand, remove the paper from between the part surface and the flat.

STEP 4   View the resultant bands from a distance of at least 10 times the diameter of the flat and at a viewing angle as perpendicular to the flat as possible (see Figure 4.13).

If the bands are straight, parallel, and evenly spaced, the surface is flat. If the bands are curved or unevenly spaced, the surface is not flat.

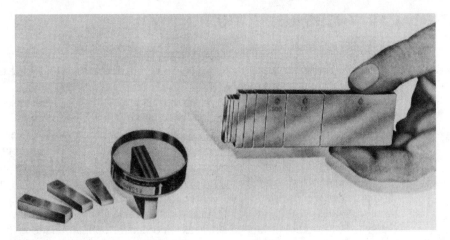

Figure 4.13 Viewing the bands on the surface of the part with an optical flat. Courtesy of MTI Corp.

Wedge Method

The wedge method is used when the surfaces of the optical flat and the workpiece are nearly parallel. The flat should be adjusted so that only four to seven bands are showing, and it should be read in two directions by changing the pressure point 90°. In cases when the wedge method is used and the part surface is convex, the bands will curve around the thin part of the wedge. When the wedge method is used and the part surface is concave, the bands will curve around the thick part of the wedge.

Contact Method

The contact method is used where the surfaces of the optical flat and the part are in intimate contact and when the wedge method cannot be used. The contact method is the best method to use when the workpiece surface is irregular, discontinuous, or ring-shaped. Using this method, the optical flat will contact on the highest point or points of the surface, and the bands will appear similar to the lines on a contour map.

If Bands Do Not Appear

1. *If bands do not appear, repeat the procedure. Do **not** slide the flat around on the part surface or the flat may be scratched.*
2. The wedge between the flat and the surface (see Figure 4.14) of the part may be too thick. If so, pressing down on the flat with even pressure may help. Again, do not slide or *wring* the flat to the part surface.

Optical Flatness Inspection   47

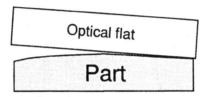

Figure 4.14 Example of the wedge method.

3. The wedge may be too thin. Moisture or oil film may cause the flat to wring to the surface (similar to a gage block) so closely that bands will not appear.
4. If the flat is at too much of an angle to the part surface, the bands may not appear. If so, try applying pressure at different points around the edge of the flat.
5. The surface irregularities may be such that the wedge method will not work. In these cases the full-contact method (see Figure 4.15) should be used.

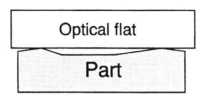

Figure 4.15 Example of the full-contact method.

Interpreting the Bands for Flatness Error

The surface is flat if straight, parallel, and evenly spaced bands appear as shown in Figure 4.16. Curved bands indicate the degree of flatness error. Figure 4.18 shows bands that represent a flatness error of about 11.57 millionths (or one band). This is measured by the imaginary line shown in Figure 4.17, where the tails of one band come in contact with the peak of the next band.

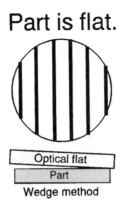

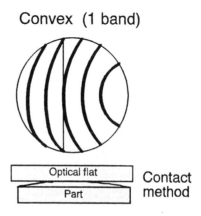

Figure 4.16 Straight, parallel, and equally spaced bands show that the surface is flat.

Figure 4.17 One band of error equals 11.57 millionths out of flatness.

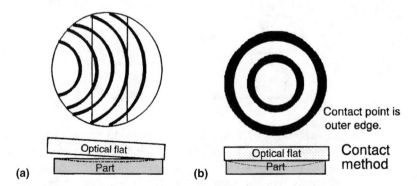

Figure 4.18 (a) Concave: out of flat by 2.2 bands (25.45 millionths). (b) Spherical concave: out of flat by 4 bands (46.28 millionths). Figures 14.17, 14.18, and 14.19 appear courtesy of Van Keuren Inc.

To assist the observer in lining up the edges of the bands, a fine piece of wire or thread could be placed on the monochromatic light's diffusion screen, or a transparent straight edge could also be used. The most frequent source of error in using optical flats is the method used (wedge or contact) and the interpretation of the bands.

The surface being inspected could be either concave or convex. Figure 4.18a and b show examples of concave surfaces using the wedge and contact methods. Figure 4.19a shows an example of convex using the contact method. The surface could also have peaks and valleys, such as the one shown in Figure 4.19b. In cases of peaks and valleys, the contact method is used, and interpretation of the error is a matter of counting the total bands. In Figure 4.19b, there are five bands total, so the flatness error is 57.85 millionths.

To summarize, flatness can be measured in millionths using optical flats properly, and, to a large extent, optical flats are functional measurements of these types of surfaces because they represent the interface of the mating part.

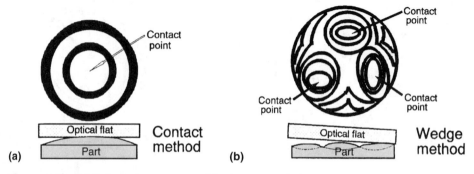

Figure 4.19 (a) Spherical convex: out of flat by two bands (23.14 millionths). (b) Peaks and valleys: out of flat by five bands (57.85 millionths). Courtesy of Van Keuren Inc.

REVIEW QUESTIONS

1. Flatness tolerances can be gaged or measured. *True* or *False?*
2. The imaginary tolerance zone for flatness of a surface is two parallel _____ the stated tolerance apart.
3. A thickness dimension to the outside of two surfaces is between .500″ and .510″. One surface for that dimension has a flatness tolerance of .003″. What is the thickest part allowable?
4. Using the jackscrew method, to achieve a faster setup, the three jacks should be _____ prior to mounting the part on them.
5. A _____ plate is used in the same manner as the jackscrew method for inspecting flatness.
6. Which of the following is an example of the fixed-plane method of flatness inspection: (a) jackscrews, (b) surface plate with built-in probe, (c) Indian pins, (d) precision parallel?
7. For surfaces that reflect light and have very tight flatness tolerances, which of the following methods would be best for flatness inspection: (a) jackscrews, (b) three equal blocks, (c) wobble plate, (d) optical flats?
8. What is another name for interference light bands?
9. Optical flat inspection methods for flatness require a _____ light source for best results.
10. Which of the following flatness inspection methods typically requires that the dial indicator (or probe) must be inverted: (a) jackscrews, (b) Indian pins, (c) optical flats, (d) precision parallels?
11. Flatness has no other datum reference except the intrinsic datum. When setting up for flatness inspection, the observer must first establish the _____ plane for measurement.
12. Name a particular gas that provides a good light source for optical flat inspection of flatness.
13. Which of the following methods is the most accurate mechanical method for inspecting flatness: (a) optical flats, (b) Indian pins, (c) jackscrews, (d) three equal blocks?
14. One flatness tolerance can be applied to control multiple interrupted surfaces for flatness. *True* or *False?*
15. Flatness tolerances cannot be used to provide more size tolerance when they are applied to surfaces that create size dimensions. *True* or *False?*

5

Straightness

OBJECTIVES

Upon completion of this chapter, the reader should be able to:

1. Understand the difference between straightness of surface elements, straightness of axis, and straightness of centerplane.
2. Understand the various tolerance zones for each straightness tolerance.
3. Measure straightness tolerances using direct or differential measurement methods.
4. Understand the virtual condition and functional gaging of certain straightness tolerances.
5. Understand and measure straightness per unit length.

INTRODUCTION AND APPLICATIONS

Straightness is a form tolerance that is not related to a datum. There are three kinds of straightness tolerances: straightness of an *axis*, straightness of *surface elements*, and straightness of a *centerplane*. Each of these tolerances is significantly different. Examples of the typical callouts for these three tolerances are shown in Figure 5.1.

As shown in Figure 5.1, we can readily see the difference between straightness of surface elements and straightness of an axis. Straightness of surface elements callouts point directly to the surface in a view on the drawing that will show the direction of surface straightness control. Straightness of an axis callouts are always next to the size tolerance of the part and have the diameter symbol in the feature control frame. The diameter symbol clarifies that the tolerance zone for straightness of an axis is a *cylinder*. Straightness of a centerplane (for noncylindrical parts) will also be called out typically

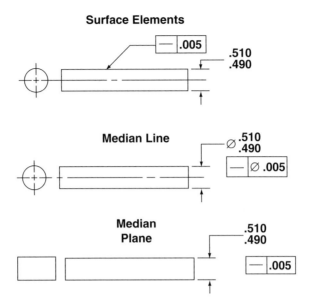

Figure 5.1 (a) Straightness of surface element requirement, (b) straightness of an axis requirement, and (c) straightness of a centerplane requirement.

under the size tolerance for the part. Figure 5.2 shows examples of the difference between straightness of an axis (or centerplane) and straightness of surface elements.

The type of straightness applied to features depends on the fit, form, and function of the part. For example, a shaft has clearance with the mating part in assembly. It is only important that the shaft goes into the mating part. Straightness of axis may be required so that the virtual condition of the shaft will be controlled (resulting in various acceptable clearances). In this case, it is not necessary to control the form of the shaft. In other cases, for example, a shaft may have a line or interference fit in assembly, and a straightness of surface elements requirement may be placed on the shaft to control the form in a manner that will provide the most contact of the surface of the shaft to the

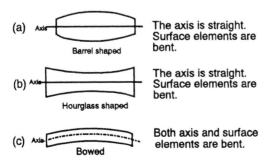

Figure 5.2 Examples of the difference between axis and surface element straightness. (a) The elements are bent, (b) the elements are bent, and (c) the axis *and* elements are bent.

surface of the hole. Straightness of surface elements applications include functional applications where barrel-shaped, hourglass-shaped, or bowed surface elements of a diameter (or surface) must be controlled.

The straightness of an axis (or a centerplane) applies to functional conditions for which the primary concern is control of the *virtual size* of the part that results from crooked axes or centerplanes. The straightness of an axis or centerplane provides for maximum use of tolerances with guaranteed fits. Another application is straightness of an axis *per unit length* for which an overall axial straightness tolerance is specified, and a usually tighter straightness tolerance is placed on a given length of the axis. This application is typically used in cases where the length of the mating part is considerably shorter than the length of the part being controlled. The tighter tolerance on axial straightness, in this case, is limited to the length of contact with the shorter mating part (where it counts).

Straightness is a unidirectional control. When applied to the surface of a noncylindrical part, it will not control flatness. Straightness, in this case, will only control the line elements of the part surface in one direction. Refer to Figure 5.3 for an example.

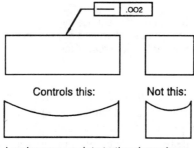

Figure 5.3 Straightness is a unidirectional control.

KEY FACTS: STRAIGHTNESS OF SURFACE ELEMENTS

1. Datum references are not applicable to any straightness tolerances.
2. The tolerance zone for straightness of surface elements is *two parallel lines* the stated tolerance apart.
3. Permissible error in the straightness of surface elements is not additive to size tolerances. The straightness tolerance must be less than the associated size tolerance. The maximum boundary of perfect form at MMC must not be violated. The worst-case condition (virtual condition) of a size feature, in this case, is the MMC size.
4. Straightness of surface elements applied to the surface of a noncylindrical part does not control flatness of the surface. The tolerance is unidirectional.
5. Straightness of surface elements is applied in design cases where size and straightness are dependent (e.g., close fits).

KEY FACTS: STRAIGHTNESS OF AN AXIS

1. The tolerance zone for straightness of an axis is a *cylinder* of the specified diameter. The diameter symbol precedes the tolerance in the feature control frame.
2. Straightness of an axis (at MMC) is the only form tolerance where bonus tolerances can be applied.
3. Straightness of an axis does not control surface elements, but straightness of surface elements does provide inherent control of the axis.
4. Straightness of an axis at MMC allows the option for functional gage design and use.
5. Straightness of an axis tolerance *per unit length* is identified by a double feature control frame where the axial straightness tolerance of the entire part is shown above and the per unit length tolerance is shown below (along with the unit length applicable).
6. Straightness of an axis (or centerplane) is applied in design cases where size and straightness are not dependent (e.g., clearance fits).

KEY FACTS: STRAIGHTNESS OF A CENTERPLANE

1. The tolerance zone for straightness of a centerplane is *two planes* the stated tolerance apart.
2. For straightness of a centerplane tolerance, the boundary of perfect form at MMC does not apply (to the extent of the straightness tolerance).
3. Straightness of a centerplane at MMC allows the option of functional gage design and use.

STRAIGHTNESS OF SURFACE ELEMENTS: CYLINDRICAL PARTS

For cylindrical parts, the feature control frame for straightness of surface elements will usually be placed in the longitudinal view of the drawing, as shown in Figure 5.4. In this case, the feature control frame states that the shaft must be straight at its surface elements within the stated straightness tolerance. The tolerance zone for this straightness requirement is *two imaginary and perfectly parallel lines that are .002" apart*, as shown in Figure 5.5.

Inspection of the part requires that the surface elements of the part be compared to an *imaginary optimum line*. Straightness of surface elements (unlike certain straightness of an axis or centerplane applications to follow) cannot be gaged; it must be inspected. The inspection methods to follow will use this figure as an example. A variety of methods can be used to measure straightness of surface elements, depending largely on the amount of tolerance allowed.

Straightness of Surface Elements: Jackscrew Method

Since the surface elements of a shaft are exposed, we can use direct measuring methods. One such method is the jackscrew method. Similar to flatness measurement

54 Chapter 5 Straightness

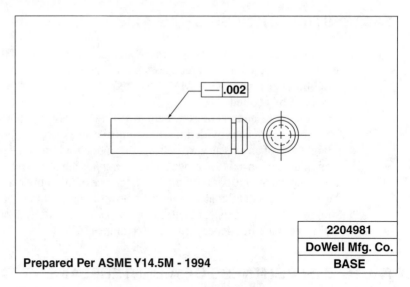

Figure 5.4 A straightness of surface elements requirement on a shaft.

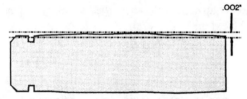

Each line element of the surface of the shaft must lie between two parallel lines .002" apart.

Figure 5.5 Tolerance zone for straightness of surface elements.

(except that we are attempting to establish an optimum *line*, instead of a *plane*), three jackscrews can be used to hold the shaft and level the surface line element to the surface plate.

Once a line element of the surface of the shaft is leveled to the surface plate, the observer can pass an indicator over the peak element of the shaft and directly measure straightness error.

The equipment needed for this method is the following:

1. Surface plate
2. Dial indicator that discriminates to less than or equal to 10% of the part straightness tolerance
3. Stand for the dial indicator (for example, surface gage)
4. Three jackscrews
5. Vee block
6. Precision parallel

Straightness of Surface Elements: Cylindrical Parts 55

STEP 1 Prelevel the three jackscrews under a standard of fixed height such as the height gage shown in Figure 5.6. Once again, preleveling the jackscrews helps save time in the setup.

Figure 5.6 Jackscrews are preleveled using a height gage.

STEP 2 Place the precision parallel over the jackscrews with two jacks at one end and one at the other. Place the part in a vee block and then the vee block on top of the jackscrews. Mark a location on one end of the part in a manner that will not damage the part. Set zero on the dial indicator at that location (rocking the indicator back and forth to find top dead center), as shown in Figure 5.7. (*Note:* Avoid using a location on the part that has an obvious flaw.)

Move the indicator to the opposite end of the part and level the part (at top dead center) using the jackscrews. (*Note:* At one end of the setup, two jackscrews are close together, and are used together to raise the part.) Take caution to ensure that the part and vee block do not fall. Mark this location also. Continue this procedure until both ends of the part (at the marked areas) are at zero on the indicator without further adjustment of the jackscrews.

STEP 3 Traverse the indicator along the surface element of the full length of the shaft and look for the full indicator movement (FIM). The FIM in this case combines all the top dead center readings on the line element of the shaft. The FIM should not exceed the straightness tolerance for the part (see Figure 5.8).

56 Chapter 5 Straightness

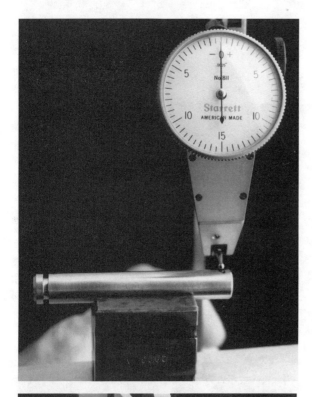

Figure 5.7 Establishing zero on the part surface at one end.

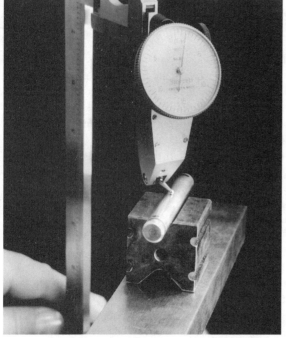

Figure 5.8 The indicator is traversed along the surface elements of the shaft (at top dead center) for straightness error measurement.

STEP 4 Rotate the part in the vee block and repeat this procedure on another line element of the shaft. It is recommended that at least four line elements (90° apart) be inspected.

In summary, the main limitation of the jackscrew method is time consumption, but the method provides very accurate results. When tight tolerances are applied to the part, the jackscrew method can be used on a wide variety of parts and lengths and is limited only by the accuracy of the dial indicator and the ability of the observer to make the setup.

Straightness of Surface Elements: Precision Straightedge Method

When tolerances are relatively loose (for example, .010″ or more), a precision straightedge and feeler wire (preferred over feeler stock) provide a faster method that can be used to inspect surface element straightness. The example tolerance of .010″ or more is based on the fact that most feeler stock is in .001″ increments, and .001″ is 10% of .010″ tolerance. Tighter tolerances could be measured with this method if the increments of feeler stock (or wire) will discriminate to 10% of the tolerance. Use of the precision straightedge provides a fast method for straightness inspection when tolerances allow. The straightedge is simply placed in contact with various surface elements of the shaft, and feeler wire is used to measure gaps. The straightedge provides a simulated straight line for reference, and the feeler wire provides measurement of straightness error, as shown in Figure 5.9.

The precision straightedge can also be used on surfaces of noncylindrical parts for measuring straightness of surface elements as long as the straightedge is placed in the one direction depicted by the drawing callout for straightness and the feeler stock used discriminates properly.

Figure 5.9 A precision straightedge can be used to check straightness of surface elements when tolerances are relatively loose.

Straightness of Surface Elements: Comparator Method

Since the actual surface elements (on cylindrical parts) can be viewed directly on an optical comparator, the comparator (and one of its crosslines) can be used to measure the part. This method is limited to the magnification power of the comparator and the ability to index the work table in discriminations that will not exceed 10% of the total straightness tolerance. Various elements of the part surface are then viewed on the comparator (see Figure 5.10) and compared using one of the crosslines as a reference (or a tolerance zone overlay) and the indexing table for direct measurement of straightness error.

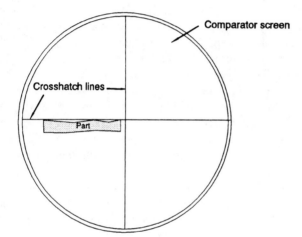

Figure 5.10 Straightness of surface elements measured on an optical comparator.

(*Note:* Straightness of surface elements is not related to a datum. Where tolerance zone overlays are used, the surface element of the part can be rotated for "best fit" within the tolerance zone lines.)

Straightness of Surface Elements: Two-Block Method

Another method that is often used for inspecting the straightness of surface elements of a cylindrical part is the use of two equal-height blocks (such as gage block stacks) to establish the *optimum line* for measurement. The method requires the following equipment:

1. Surface plate
2. Dial indicator of appropriate discrimination (10% of the straightness tolerance)
3. Surface gage (or height gage) to mount the dial indicator
4. Two equal-height gage block stacks (or precision parallels)

STEP 1 Place the two equal blocks on the surface plate at a distance apart that will contact each end of the part, and place the part on the blocks as shown in Figure 5.11. (*Caution:* Be careful not to let the part roll from the blocks. As shown in the figure, other gage blocks can be used to stop the part from rolling. Stabilize the part in a way that stops rolling but does not raise the part from the blocks or interfere with the underside measurement of line elements.)

Figure 5.11 The part is placed on top of two equal-height blocks.

STEP 2 After stabilizing the part, indicate on the underside the line element that is in direct contact with the blocks, as shown in Figure 5.12. Be sure to indicate only the bottom dead center of the line element across the full length of the part.

STEP 3 Repeat the above procedure on at least three other line elements by rotating the part 90° (recommended).

The two-equal-block approach is a quick method for inspecting straightness of surface elements as long as the blocks are equal height within gage tolerances and the part does not roll or rise up from the blocks during the measurement.

A limitation to this method is that the areas at each end of the part cannot be indicated because they are resting on the blocks.

STRAIGHTNESS OF SURFACE ELEMENTS: NONCYLINDRICAL PARTS

As shown previously in Figure 5.3, straightness of surface elements on noncylindrical parts applies only in one direction. The applicable direction is identified by the particular view of the drawing where the leader arrow points.

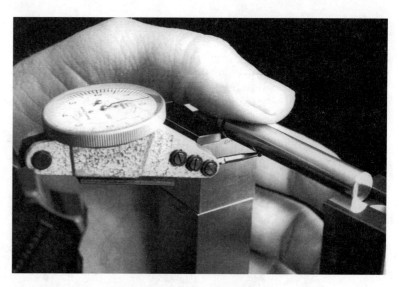

Figure 5.12 Dial indicator shows line elements of the surface.

Straightness of surface elements on noncylindrical parts can be measured using the jackscrew method or precision straightedge method previously covered, but the optical comparator cannot be used visually since a line element of the noncylindrical surface cannot be seen. A comparator could be used for straightness of surface elements on a noncylindrical part if the comparator were equipped with (1) a *tracer element* with a reticle follower shown on screen and (2) an *overlay* that has two lines that are the straightness tolerance apart (accounting for the magnification power of the comparator). In this manner, the tracer could follow a line element of the part surface and the reticle follower could be viewed on the comparator screen with respect to the two lines of tolerance on the overlay.

STRAIGHTNESS OF AN AXIS

Applications

There are not many applications for axial straightness control because the tolerance applies only to the axis itself and it creates a virtual condition (outer boundary) beyond size limits. Straightness of an axis is applied where the size of a feature (such as a shaft) is independent from the straightness of that feature. Examples include mating features where there is plenty of clearance and long-size features with limited interface. For most applications, straightness of surface elements is a better alternative. In fact, straightness of surface elements automatically controls the straightness of an axis.

Straightness of an axis (or centerplane) tolerance is concerned only with the straightness of the axis (or centerplane), not the surface elements. This particular tolerance is also the *only* form tolerance that can be gaged (when applied at MMC).

In this case, per rule 3 of the standard, the MMC modifier must be stated in the feature control frame in order for functional gaging to apply. A functional gage is designed only to gage the *virtual condition* of the part. (There will be more on axial straightness at MMC later in this chapter.) Feature control frames for straightness of an axis (or centerplane) are usually called out under the size tolerance, and in the case of axial straightness the callout will have a diameter symbol preceding the tolerance. The diameter symbol clarifies the fact that the tolerance zone is an *imaginary cylinder* of tolerance in which the axis must lie. Straightness of an axis is more difficult to measure than surface element straightness because the observer must prove that the axis lies within the cylindrical tolerance zone.

Figure 5.13 shows an example of a part with straightness of an axis tolerance (due to rule 3, RFS is understood in this requirement). The feature control frame states that the axis of the diameter must be straight within a .002″ diameter. The tolerance zone for the part in Figure 5.13 is an *imaginary cylinder of .002″ diameter within which the axis of the controlled feature must lie*. This tolerance zone is shown in Figure 5.14. Figure 5.15 shows the same part drawing, except that MMC has been specified in the feature control frame. In this MMC case, a functional gage can be used. The tolerance zone is still an imaginary cylinder, but when MMC is specified, it is only a .002″ diameter cylinder when the diameter being controlled is produced at MMC size. If the diameter is produced, in this case, smaller than MMC size, a bonus axial straightness tolerance is allowed (as shown in Figure 5.16).

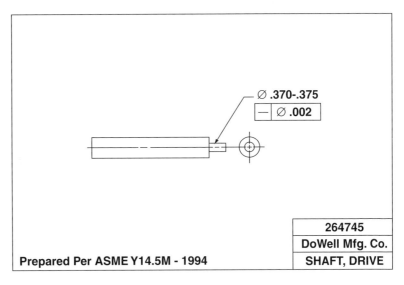

Figure 5.13 Drawing requirement for straightness of an axis at RFS.

Chapter 5 Straightness

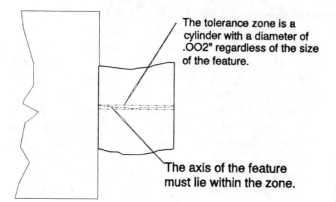

Figure 5.14 Tolerance zone for the part.

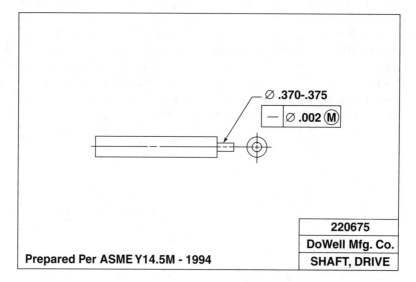

Figure 5.15 Drawing requirement for straightness of an axis at MMC.

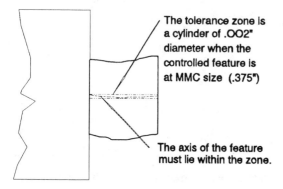

Actual Feature Size (")	Tolerance Zone Diameter (")
.375	.002"
.374	.003"
.373	.004"
.372	.005"
.371	.006"
.370	.007"

Figure 5.16 Tolerance zone for the part.

STRAIGHTNESS OF A CENTERPLANE

For noncylindrical parts, such as the part shown in Figure 5.17, the drawing callout for straightness of the centerplane of the part will still be located under the size tolerance on the drawing. The feature control frame states that the centerplane of the part must be straight within .002″. The tolerance requires that the imaginary centerplane of the part must be straight within a tolerance zone of .002″ *(two parallel planes .002″ apart in which the centerplane of the part must lie)*, as shown in Figure 5.18. (Straightness of a centerplane will be covered later in this chapter.)

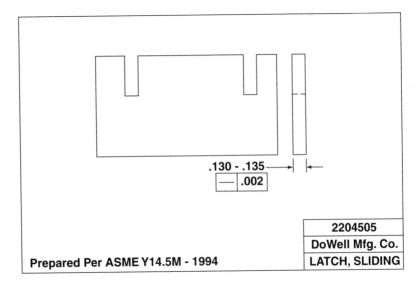

Figure 5.17 Drawing requirement for straightness of a centerplane.

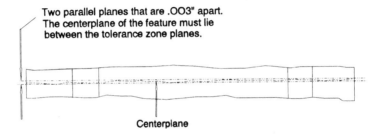

Figure 5.18 Tolerance zone for the part.

INTRODUCTION TO FUNCTIONAL GAGING: STRAIGHTNESS EXAMPLES

Functional gaging is allowed in cases when features of size are controlled with a geometric tolerance at MMC. When MMC is stated in the feature control frame, bonus tolerances on the control applied can be used as the actual size of the controlled feature departs from its MMC size toward LMC.

The reason that bonus tolerances are allowed is that the *virtual condition (or worst-case allowed)* never changes. An example drawing and table of bonus tolerance allowed is shown in Figure 5.19. (*Note:* When RFS applies, bonus tolerances are not allowed. When LMC applies, bonus tolerances are allowed, but functional gaging cannot be used.)

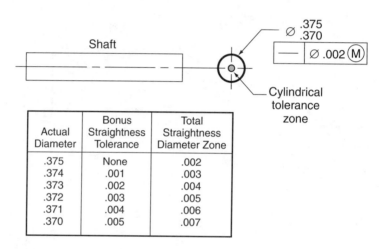

Figure 5.19 Example drawing and allowable bonus tolerances dependent on feature size.

Figure 5.19 shows that if the shaft is produced at MMC size (.375″), no bonus tolerance is allowed. The shaft must be straight within the specified straightness tolerance of .002″. If the shaft is produced at .373″ diameter (.002″ smaller than MMC), the .002″ can be added to the stated .002″ tolerance for straightness, allowing a total of .004″ straightness error.

Once again, the reason for allowing the bonus tolerance is the fact that the *actual virtual size never violates the allowed virtual condition*. The virtual condition of this shaft is the MMC limit (.375″ diameter) plus the stated straightness tolerance (.002″), which equals .377″. In all the cases shown in Figure 5.19, the actual diameter plus the total straightness tolerance (including the bonus) equals .377″. In this way, design engineers can allow more tolerance, depending on size, and still guarantee that the mating parts will assemble with the appropriate clearance or interference. When a geometric tolerance is modified with MMC in the feature control frame, functional gages can be designed and used at the discretion of the producer.

A functional gage is a design of hard tooling that "functionally" gages geometric worst-case conditions. Functional gages represent a *worst-case mating part* because they are designed to gage the virtual condition of the part. The *virtual condition* is the collective effect of size and allowable geometric errors in a part. For example, a shaft that has a size tolerance of between .500″ and .510″ and a straightness tolerance of .002″ at MMC has a virtual condition. The drawing allows the shaft to be as large as .510″ diameter, *and* it can be bent by .002″ when it is at .510″ diameter. Therefore, the virtual condition is .510″ diameter *plus* .002″ straightness error equals .512″.

If this design required clearance between the shaft and the hole, the clearance would be based on the worst-case (virtual) condition of the shaft (.512″), and the virtual condition of the hole would be designed for the required clearance. For a hole, the virtual condition is equal to the MMC size *minus* the amount of geometric error allowed. For example, the hole in the mating part for the shaft discussed earlier has a size tolerance of between .514″ and .515″ and a straightness tolerance of .001″. In this case, the *virtual condition* of the hole is MMC (.514″) minus the .001″ straightness tolerance, which equals .513″. Therefore, the designed clearance between the shaft virtual condition (.512″) and hole virtual condition (.513″) is .001″. The following equation will be helpful for calculating the virtual conditions of external features (for example, shafts, pins, and bosses) and internal features (for example, holes, slots, and grooves):

$$\text{Virtual condition}_{(external\ features)} = \text{MMC} + \text{Geometric tolerance}$$
$$\text{Virtual condition}_{(external\ features)} = \text{MMC} - \text{Geometric tolerance}$$

For the shaft in question, the design of a functional gage is relatively simple. To functional-gage the shaft (to represent the mating part), a ring gage can be designed that will represent the worst case of the hole in the mating part. The ring gage must be designed to (1) be long enough to encompass the entire length of the shaft, and (2) have an inside diameter that is based on the virtual condition of the shaft. The gage tolerances should not consume more than 10% or less of the shaft tolerance. A picture of the functional gage for the part in Figure 5.19 is shown in Figure 5.20. (For more detailed information on functional gages, refer to Chapter 18.)

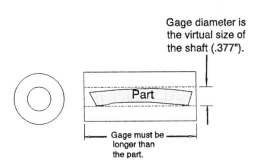

Figure 5.20 Functional ring gage for straightness of a shaft at MMC.

STRAIGHTNESS OF AN AXIS: RFS

One of the more complex form tolerances is straightness of an axis RFS. When RFS is specified, the axial straightness applies regardless of the size of the part. Therefore, there are several allowable virtual conditions, and functional gaging cannot be used. The straightness of the axis must be inspected. For cylindrical parts, the tolerance zone for axial straightness is an *imaginary cylinder* in which the axis of the part must lie. For noncylindrical parts, the tolerance zone for centerplane straightness is *two parallel planes* the stated straightness tolerance apart. In both cases, if MMC modifies the tolerance, a functional gage can be used to gage the part for straightness. But if RFS modifies the tolerance, the axial or centerplane straightness must be measured.

Straightness of an Axis: Differential Measurement Method

One method of finding the straightness error of a shaft (without advanced measuring equipment) is to explore the curvature of opposing elements in order to find out the error of the axis that lies in between. This method is time-consuming but often yields good results. Many straightness-of-axis errors are somewhat obvious once the elements have been explored. To begin the discussion by viewing elements, refer to Figure 5.21 for examples of possible axial errors.

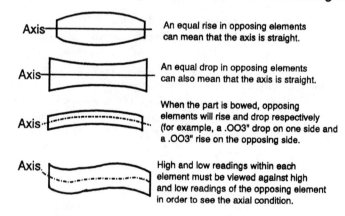

Figure 5.21 Exploring opposing elements of a shaft to measure axial straightness error.

The equipment required for this method is as follows:

1. Surface plate
2. Dial indicator and stand (such as a surface or height gage)
3. Pencil and grid paper

STEP 1 Gently roll the shaft to be measured onto the surface plate.

STEP 2 Pass the dial indicator across the top line element of the shaft, rolling it slightly after each pass. The purpose of this is to find the high spot of the shaft with reference to the surface plate. Once an extreme area (for example, high spot) is found, mark the top of the shaft (in a manner that will not damage the part) and make sure that the shaft is stabilized so that it will not move. Avoid clamping the shaft on the top surface.

STEP 3 Using grid paper and pencil, explore the entire line element, recording high and low values from one end of the element to the other. On the paper, record the locations (for example, 1″ from the end) of high or low areas. Make sure that the indicator is rocked back and forth to read only the peak of the line element during the measurement, as shown in Figure 5.22.

STEP 4 Once this is completed, using the mark on the shaft, rotate the shaft 180° (the marked area will now be touching the surface plate) so that the direct opposing elements can be explored.

STEP 5 Repeat step 3, recording the locations, amount, and direction of the high and low areas of the opposing line element.

STEP 6 Review the graph and draw a sketch of the elements of the shaft, such as the one in Figure 5.23.

As shown in Figure 5.23, both opposing elements of the shaft depicted a barrel-shaped condition due to both opposite elements having high and low areas in the same location. In a symmetrical barrel-shaped condition, the axis is straight and the surface elements are bent. The part is fairly symmetrical; therefore, the axis is straight and the

Figure 5.22 Exploring a line element of the shaft.

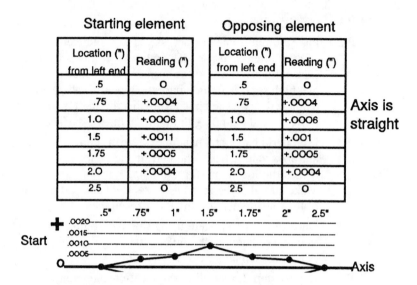

Figure 5.23 Graph of the surface elements of the shaft.

line elements are bent in excess of .001″ each. The requirement, however, is axial straightness; therefore, the observer would accept this part.

Measurement of axial locations or form at RFS is a complicated procedure in open setup (or any setup for that matter). The most significant limitations to the differential method are the following:

1. Out of roundness of the shaft can contribute to error in the results.
2. The observer must be able to graph line elements and interpret the *differential* values of opposing elements.
3. The differential method is very time-consuming.

There are, however, more sophisticated measuring instruments today that make this task easier (such as the precision spindle method to follow), but the evaluation of the results can still be a complicated matter.

Straightness of an Axis: Precision Spindle Method

To track the axis of a diameter (which is the center of all circular sections of the diameter), we must be able to find the axis regardless of the roundness error of each circular section. A precision spindle (see Figure 5.24) is a device that rotates true within tight tolerances and has a probe or an indicator that is able to evaluate and graph the actual diameter sections of the shaft. The spindle, when accompanied by a precision vertical slide, can give the observer the ability to evaluate all surface elements of the shaft, graph the results, and analyze axial straightness error.

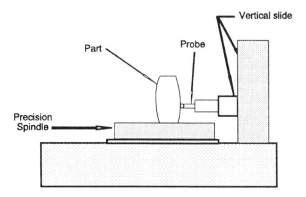

Figure 5.24 Sketch of a precision spindle and vertical slide.

Precision spindles with mechanical bearings can rotate true within tight tolerances such as .00005″, whereas spindles with air bearings can achieve .000005″ true rotation. Hence, the spindle provides a perfect circle for comparison.

Precision vertical slides such as a dovetail slide (or air-bearing slide) provide a path for the probe to sweep perfectly parallel (within gaging tolerances) to the axis of the spindle. When these two devices are used together and the probe sends signals to a computer to graph the results, various complicated measurements such as axial straightness, roundness, concentricity, and cylindricity become easier to inspect.

Figure 5.25 shows an example of a polar graph produced by the precision spindle method. In this graph, we can see the error of the axis along various circular sections of the part. To produce this graph, we must set the probe at one end of the shaft and evaluate various sections separately.

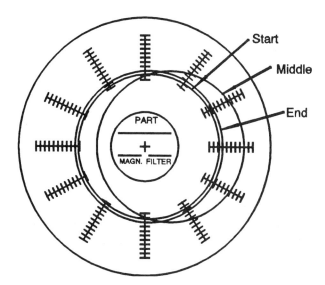

Figure 5.25 Example of a graph for axial straightness error.

70   Chapter 5   Straightness

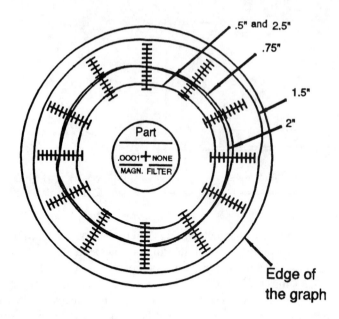

Figure 5.26 Polar graph for axial straightness shows the part is acceptable.

Using the barrel-shaped part that was measured in the previous differential method, the precision spindle and slide measurements (in the same locations) will produce a polar graph similar to Figure 5.26.

Figure 5.26 verifies that the differential method, in this case, was correct. The axis of the part is straight well within the axial straightness tolerance. In Figure 5.26, the observer can see the barrel-shaped condition, knowing that the circular graph at 0.5" from the end and the graph at 2.5" from the end are in the same location, and the middle graph at 1.5" from the end is symmetrically wider. Interpreting the polar graphs requires a certain level of practice.

Some examples of different part conditions on polar graphs are shown in Figure 5.27 to help the observer understand the various problems that can be seen with a precision spindle and vertical slide.

Straightness of axis measurement requires an inspection method that will evaluate the axis of the diameter with respect to a cylindrical tolerance zone. The inspection method and results must be compared to the cylinder of tolerance.

STRAIGHTNESS OF A CENTERPLANE AT MMC

For noncylindrical parts or part size features (for example, the center of slots, bosses, or square tabs), straightness of the centerplane is evaluated. Since there is no axis, the imaginary tolerance zone for straightness of a centerplane is *two parallel planes* that are the stated tolerance apart. When MMC applies, a bonus tolerance is allowed if the feature of size departs from MMC toward LMC. If RFS applies, the straightness tolerance is fixed (regardless of feature size).

Straightness of a Centerplane at MMC

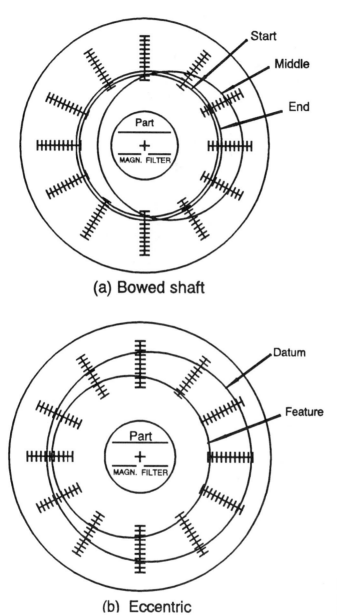

Figure 5.27 Various polar graph results indicating geometric error.

Review the drawing in Figure 5.28. A straightness tolerance has been applied to the thickness of the part. Therefore, the centerplane of the part must be straight within .002″ when the part is at MMC size (.135″). In this case, bonus tolerances apply and a functional gage can be used. In lieu of a functional gage, functional gaging can be simulated with surface plate accessories.

Chapter 5 Straightness

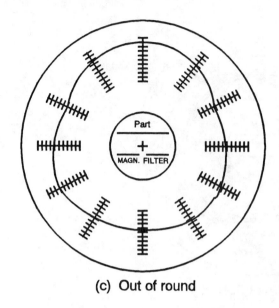

(c) Out of round

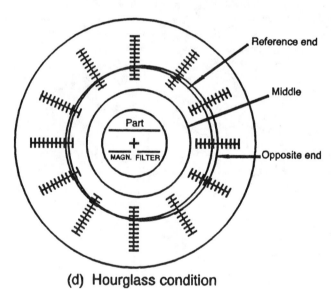

(d) Hourglass condition

Figure 5.27 (Continued)

Functional Gaging Straightness of a Centerplane at MMC

A functional gage for the part in Figure 5.28 is relatively simple. Some basic requirements of the gage design include the following:

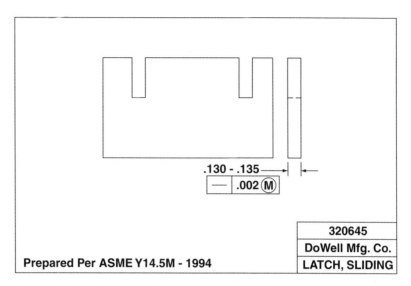

Figure 5.28 Straightness of a centerplane at MMC.

1. The gage must have two plates within gaging tolerance that are large enough to surround the part on both sides.
2. The plates must be separated in a manner that will simulate two planes that are the virtual condition of the part (.137″).

Figure 5.29 is a sketch of the functional gage and part. In this case, if the part passes through the functional gage without resistance, the virtual condition has been met, and the only thing left to do is measure the thickness tolerance of .130″ to .135″. (*Caution:* Functional gages measure only the virtual condition. The part could be oversize with a straight centerplane, and the functional gage will not see this condition. We must always measure or gage size separately from virtual size.)

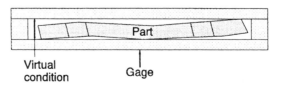

Figure 5.29 The part is introduced to the functional gage for straightness at MMC.

Simulating a Functional Gage for Straightness of Centerplane at MMC

If it is not desirable to design and build a functional gage, the part can still be functionally inspected using surface plate accessories. The equipment required for this inspection is as follows:

1. Two gage block stacks of .137" dimension (the virtual condition of the part)
2. Two precision parallels

STEP 1 Make sure the gage blocks are stacked at .137" dimension by using a micrometer as a cross check.

STEP 2 Place the .137" gage block stacks between two precision parallels. (*Note:* The parallels must be wide enough to encompass the entire part.)

STEP 3 Introduce the part between the parallels without applying pressure, as shown in Figure 5.30. Notice in Figure 5.30 that the part is completely encompassed between the parallels. This is necessary for functional gaging. If the part passes through the parallels and, at one point, the part is completely enclosed by the parallels, it is straight within .002" at MMC. Once that is completed, check the thickness tolerance (.130" to .135") with a micrometer.

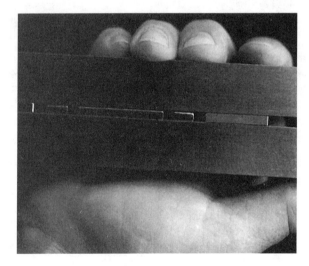

Figure 5.30 Simulating a functional gage with precision parallels and gage block stacks.

This example of using surface plate accessories to simulate a functional gage is a very practical method of functional inspection when hard tooling is not affordable or desirable.

STRAIGHTNESS OF A CENTERPLANE (RFS)

When a straightness tolerance of a noncylindrical part is stated at RFS (regardless of feature size), functional gaging cannot be used. The observer must verify that the centerplane is within an imaginary tolerance zone of *two parallel planes the stated tolerance apart*. Also, where RFS applies, no bonus tolerance can be allowed. The part must be straight within the stated tolerance regardless of size. For this example, the previous part drawing will be changed to RFS, as shown in Figure 5.31.

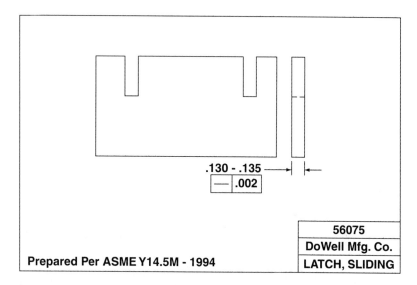

Figure 5.31 Straightness of a centerplane (RFS).

The tolerance zone for straightness of a centerplane is *two imaginary planes the stated tolerance apart within which the centerplane of the controlled feature must lie*. In the drawing in Figure 5.31, these planes are fixed at the stated tolerance due to the understood RFS (general rule 3). Refer to Figure 5.32 for an example of the shape of the tolerance zone.

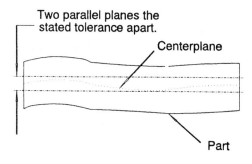

Figure 5.32 Tolerance zone for straightness of a centerplane.

Straightness of a Centerplane: Differential Measurement Method

Since this type of straightness requirement cannot be gaged (due to the RFS modifier), it must be inspected in an open setup. Finding the straightness of a centerplane can be as complicated in open setup as the axial straightness covered earlier in this chapter. In this case, as with axial straightness, differential methods must also be used. The equipment required for this method is the following:

1. Surface plate
2. Dial indicator with discrimination to no more than 10% of the straightness tolerance
3. Height gage or surface gage to hold the indicator
4. One or two precision parallels
5. Two equal-height gage block stacks
6. Pencil and pad of grid paper

STEP 1 Lay the part gently on the two gage blocks where the contact is made at the extreme ends of the part. Take caution not to drop the part. Find a way to apply pressure at each end of the part just above the gage block stack so that the dial indicator will not raise the part (this is especially true for small parts).

STEP 2 Using the precision parallel(s) as a guide for the base of the height gage (or surface gage), the dial indicator can be traversed on the underside of the part in a straight line. Beginning in one corner, set the indicator at zero and traverse the indicator across a line element, recording the high and low areas. Repeat this step on at least three line elements, making sure that the guides are reset so that each pass is in a straight line, as shown in Figure 5.33.

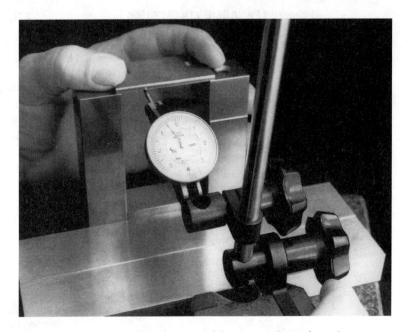

Figure 5.33 Indicating line elements of the part on the surface plate.

Straightness of a Centerplane (RFS) 77

STEP 3 Lay the part on its opposite side on the surface plate, remembering the line elements previously measured.

STEP 4 Set zero on the indicator again and traverse the opposing elements, recording the high and low spots. Refer to Figure 5.33 again since the part being used is symmetrical. On this opposite side, try to indicate the line elements in the same locations as previously indicated.

STEP 5 Evaluate the initial results of the differential readings.

In evaluating the results of the differential measurements made on the part, there can be a variety of possible outcomes, as shown in Figure 5.34. The conditions covered previously on straightness of an axis measurement could also occur on the centerplane. The part could be barrel-shaped, hourglass-shaped, or tapered. All these conditions will cause the differential readings to be about the same amount, but in different directions. For example, if the part is barrel-shaped, the differential readings will be about the same amount, but the middle opposing readings will both be plus. If the part is hourglass-shaped, the differential reading will also be about the same amount, but the middle of both opposing line elements will be minus. If the part is just tapered, all readings of line elements will trend downward from one end to the other.

Noncylindrical parts: Straightness of Centerplanes		Readings			Results
		Start	Middle	End	
Start 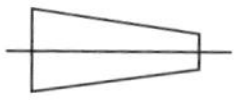 — Centerplane Tapered	Start elements Opposite elements	Plus Plus	Minus Minus	Minus Minus	Centerpiece is straight
Start  — Centerplane Barrel shaped	Start elements Opposite elements	Minus Minus	Plus Plus	Minus Minus	Centerpiece is straight
Start  — Centerplane Hourglass shaped	Start elements Opposite elements	Plus Plus	Minus Minus	Plus Plus	Centerpiece is straight
Start  — Centerplane Bowed	Start elements Opposite elements	Minus Plus	Plus Minus	Minus Plus	Centerpiece is crooked

Figure 5.34 Some possible outcomes of centerplane straightness measurement.

In the case of barrel-shaped, hourglass-shaped, or tapered parts, the centerplanes are all straight. Only in cases where the part is bowed or S-shaped, for example, will the centerplane be crooked. If the part is bowed, the line elements on one side (from the zero setting) may indicate zero, plus, then 0. Then the opposing line elements would indicate 0, minus, then 0. In other words, the middle readings of opposing line elements will have

Chapter 5 Straightness

opposite directions. In cases of S-shaped parts, one side line element may indicate (from zero setting) plus, minus, plus; then the opposing line element (from zero setting) would indicate the opposite (minus, plus, minus).

REVIEW QUESTIONS

1. It is possible for a part to have a diameter where the surface elements are crooked, but the axis is perfectly straight. *True* or *False?*
2. Keeping the general rules in mind, what is the virtual size of a shaft that has a size tolerance of between .750" and .755" diameter and a straightness of surface elements tolerance of .002"?
3. The tolerance zone for straightness of surface elements of a cylindrical size feature is two perfectly parallel _____ the stated tolerance apart.
4. The tolerance zone for straightness of an axis is a _____.
5. The tolerance zone for straightness of a centerplane is _____.
6. The diameter of a shaft has a size tolerance of between .750" and .760" and a straightness of axis tolerance of .005" diameter. What is the virtual condition of the shaft?
7. If the shaft in Question 6 had an actual diameter of .754" and the axis were out of straightness by exactly .002", what is the actual virtual size of the shaft?
8. When using two jackscrews to inspect straightness of surface elements, the jackscrews help the observer to establish the optimum _____ for measurement.
9. The most effective methods for measuring straightness of an axis (regardless of feature size) use _____ measurements to track the axis.
10. When the surface elements of a cylindrical feature are being controlled, an optical comparator can be used to measure surface element straightness. *True* or *False?*
11. A noncylindrical part must be functionally gaged for straightness of its centerplane at MMC. A hard gage could be used, but cost is a problem. Can the observer use surface plate accessories (such as gage blocks or precision parallels) to set up a virtual boundary for worst-case gaging of the part? *Yes* or *No?*
12. A shaft has a size tolerance of between .500" and .505" and a straightness of axis tolerance of .002" at MMC. A functional ring gage design has been chosen for the functional gage. What is the nominal size of the hole in the gage (before allowing for gage-makers' tolerance and wear allowance) to functionally gage this part?
13. When straightness of surface elements tolerance is applied to surfaces of noncylindrical parts, the straightness tolerance only applies in the direction shown by the drawing. *True* or *False?*
14. With looser straightness-of-surface element tolerances, feeler wire and a precision straight _____ can be used to measure straightness error.
15. Straightness-of-surface elements automatically control the straightness of an axis. *True* or *False?*

6

Circularity

OBJECTIVES

Upon completion of this chapter, the reader should be able to:

1. Understand circularity tolerances and tolerance zones.
2. Identify the various lobes of circularity error.
3. Understand that two-point measuring instruments are error-prone when used for measuring circularity.
4. Know the difference between measuring circularity and verifying circularity.
5. Apply various direct measurements of circularity, such as vee-block measurements, air gages, and intermicrometers.
6. Understand that size is adversely affected by circularity error.
7. Understand precision spindle methods and the use of polar graphs to measure circularity error.

INTRODUCTION AND APPLICATIONS

Circularity is a form tolerance that applies to cylindrical parts. Circularity tolerances are specified in cases when the function of the part necessitates that each circular element be round regardless of size or when size tolerances (per rule 1) do not provide sufficient control of the circularity of the part.

Circularity is also used, at times, to qualify a datum diameter. It is often more practical to apply circularity tolerances to parts that have short axes (such as gaskets, washers, and short sleeves) to control the circularity at each circular element and to control the effective size of the part. An example is the inside diameter of a ring for which the function is to slide along a shaft. As with other form tolerances, circularity is not specified to

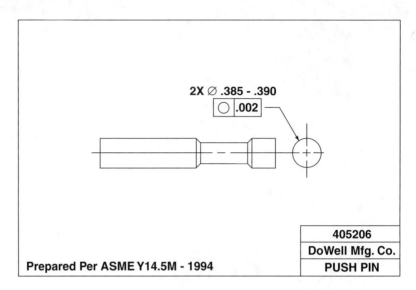

Figure 6.1 Circularity requirement on a drawing.

a datum. Another important thing to know about circularity is that the tolerance zone is *radial,* not diametral. Circularity (where rule 1 of the standard applies) is automatically required by virtue of the size tolerances on a part diameter unless a specific feature control frame identifies a circularity tolerance for that diameter that refines the inherent circularity control provided by size limits. Figure 6.1 is an example of a drawing callout for circularity. The feature control frame states that the diameter shall be round within the stated tolerance on radius. The tolerance zone for circularity is *two concentric circles of which the radii are the stated tolerance apart.* Figure 6.2 is an example of the tolerance zone.

KEY FACTS

1. Circularity tolerance (per ASME Y14.5) is a radial tolerance zone, not diametral.
2. Circularity cannot be gaged; it must be measured.
3. Size tolerances (per rule 1) control circularity within size boundaries unless a specific circularity control has been applied to the size feature that refines the size tolerance.
4. Circularity tolerance cannot violate the maximum boundary of perfect form of the size tolerance.
5. Circularity tolerance is not associated with a datum reference.
6. Circularity is regardless of feature size. Size may vary within size limits, but the circularity tolerance remains the same in any case.
7. The circularity tolerance applies at each circular section of the diameter independently.
8. Circularity tolerances can be applied to conical parts since individual circular sections must be round within tolerances.

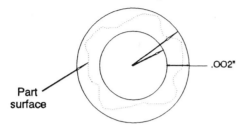

Figure 6.2 Circularity tolerance zone for the part.

THE PROBLEM OF LOBES

One of the most significant problems with respect to circularity measurement is the fact that most products produced are *not* simply oval-shaped to the point where circularity error can be measured easily with standard two-point measuring instruments, such as micrometers, calipers, indicating snap gages, and other similar instruments. When parts are not oval-shaped, two-point measurements are always in error when it comes to evaluating circularity. The main cause of errors when measuring circularity with two-point instruments is the *lobes* of the part.

Two primary errors are made when using two-point measuring devices such as the ones previously mentioned. For one, the device cannot directly measure the *effective diameter* (which is the diameter on a shaft, for example, that would come in contact with the smallest cylinder placed over the shaft). The effective outside diameter, due to circularity error, is larger than the distance measured by a two-point device. The other error is caused by odd, equally spaced lobes on the part that cannot be seen by two-point instruments because the lobes (and the height of each) are transparent to two-point instruments.

Therefore, two-point measuring instruments give the observer little or no information about the size or circularity of the part. For example, the oval-shaped part (two lobes) shown in Figure 6.3 is the only configuration for which a two-point measuring device could be used to measure circularity. The radial circularity error, in the case of the oval part, would be one-half the difference between the largest and the smallest diameter measurements. The trilobe, five-lobe, and higher odd-lobed parts (as shown in Figure 6.3) will fool the observer every time when micrometers, calipers, and other two-point instruments are used (as shown in Figures 6.4 and 6.5). With these types of odd-lobed conditions, the part will actually be larger than the measured diameter, and the circularity error will be worse. In fact, odd-lobed parts usually appear perfectly round when measured with two-point instruments.

THE EFFECTIVE SIZE

The effective size of a part (due to circularity error) is the diameter at the extremes of the lobes. A simple example is an oval-shaped outside diameter on a part. The effective

Chapter 6 Circularity

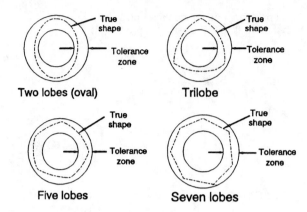

Figure 6.3 Various lobes of circularity error.

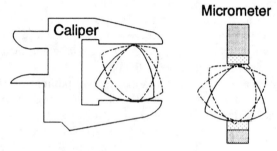

Figure 6.4 Parts appear round although they are not, when using two-point measuring instruments.

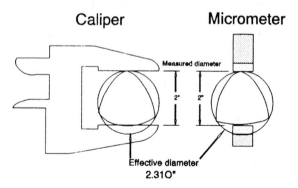

Figure 6.5 Two-point gages cannot measure effective size when parts have odd-shaped lobes.

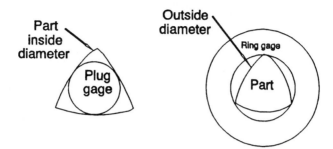

Figure 6.6 Plug and ring gages detect the effective size of lobed diameters.

size of an oval-shaped outside diameter is the smallest true circle that will fit over the diameter (touching the peaks of the two symmetrical lobes). The effective size of odd-lobed outside diameters such as the trilobe part in Figure 6.6 can be quickly seen when a ring gage (or plug gage) is used.

The ring (or plug) gage is perfectly round (within gage tolerances) and, therefore, makes contact with the extreme conditions (peaks of the lobes) of the part. To measure the effective size of an odd-lobed part, radial measurements must be made on the part to the point where the number of lobes and the height of the lobes are known. A precision spindle can be used for this purpose (discussed later in this chapter). Two-point measuring instruments do not suffice for diameter measurements where equally spaced odd lobes exist. Figure 6.6 shows how ring gages and plug gages detect the effective size of outside and inside diameters, respectively, that have odd-shaped lobes.

VEE-BLOCK METHOD: OUTSIDE CIRCULARITY MEASUREMENT

Since two-point measuring instruments are not recommended for parts with odd lobes and many companies may decide not to purchase a precision spindle, other open setup techniques can be used to measure circularity. One effective circularity measurement method for outside diameters is the vee-block method. This method provides three points of contact on the diameter (two points in the vee block and one point at the indicator). Although there are limitations (and a certain amount of measurement error), the vee-block method will show more accurate results than any two-point instrument.

The primary concern when using the vee-block method is that the vee block must be a specific angle in order to effectively measure circularity error *and* the FIM results require conversion to obtain the *actual radial circularity error of the part*. The angle of the vee block is calculated using the equation in Figure 6.7.

For example, a trilobed part (three equally spaced lobes) would require a 60° vee block for circularity measurement since $n = 3$, 360° divided by 3 equals 120°, and 180° minus 120° equals 60°. This is a limitation (when using the vee-block method) because the actual number of lobes on the part is unknown unless an instrument like the precision spindle is used. For most applications, a 60° vee block will usually suffice. A secondary

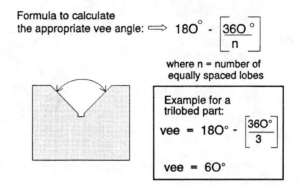

Figure 6.7 Proper angle of a vee block for circularity measurement.

limitation of the vee-block method is that the method, with the correct-angle vee block, can exaggerate the actual circularity error of the part.

For this reason, the observer must compensate for this error in the results. The table of constants in Figure 6.8a will help compensate for the FIM values found for specific lobes and vee-block angles, but the observer can use the equation in Figure 6.8b for these and other cases. It should also be understood that the vee-block method is *not* recommended for even-lobed parts such as those which are oval-shaped.

The error associated with a constant 60° vee block exists and can be significant, but it is negligible compared to the error of two-point instruments. Another limitation is that circularity applies to each circular section of the diameter; therefore, a knife-edge vee block is preferred so that only a circular section is contacted. Knowing these limitations, the vee-block method of circularity measurement is still a considerably more effective method than any two-point measuring instrument. It is highly recommended that the observer obtain, at least by sampling, the lobes of the part through the use of a precision spindle. One way this can be done, if a precision spindle is not owned by the manufacturer, is to send sample parts from the process to a measurement laboratory (or another manufacturer who owns the equipment) to obtain lobe information.

Another way is to try various vee blocks of different angles to find the one block angle that shows the highest (worst) indications; then use that vee block and the proper correction factor to obtain the results needed.

Another option is to use a rotary table to obtain lobe information about the part and then select the proper vee block. No single vee-block angle will suffice for all lobes.

Consider the part in Figure 6.9. The following equipment is necessary for the measurement:

1. Surface plate (or bench comparator)
2. Dial indicator that discriminates to 10% of the circularity tolerance
3. Vee block large enough to hold the part at the correct angle (60° if the number of lobes is unknown)
4. Surface gage, height gage, or stand for the indicator

Vee-Block Method: Outside Circularity Measurement

Number of Equally Spaced Lobes	Angle of the Vee Block	Divide the FIM Result by:	Example: FIM Value Found	Example: Actual Roundness Error
3	60°	3.000	.003"	.001"
5	108°	2.236	.003"	.0013"
7	128° 34'	2.110	.005"	.0024"
9	140°	2.064	.004"	.0019"

(a)

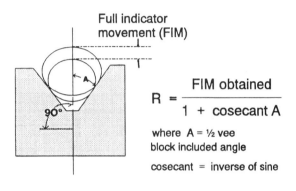

$$R = \frac{\text{FIM obtained}}{1 + \text{cosecant } A}$$

where A = ½ vee block included angle

cosecant = inverse of sine

(b)

Figure 6.8 (a) A table of conversion factors for vee-block circularity measurements, and (b) an equation for converting FIM results for specifically lobed parts when appropriate vee-block angles are used.

STEP 1 Place the vee block on the surface plate or bench comparator, and place the part in the vee block.

STEP 2 Place the indicator at top dead center (directly between the contact areas of the vee block) on the part on one section of the diameter, and rotate the part 180° as shown in Figure 6.10. Look for the FIM value during rotation. Divide the FIM value by the proper conversion factor (see Figure 6.8), or divide by 3 if lobes are unknown. (*Note:* Appreciable error can result by using a constant 60° vee block when lobes are unknown. Once again, it is recommended that the product be radially measured to establish the number and location of lobes, at least on a sample. After lobe information has been obtained, the correct-angle vee block and conversion factors can be used to measure circularity.)

STEP 3 Move the indicator to another circular section of the part and repeat step 2. If the FIM at all circular sections is within the circularity tolerance, the part is acceptable.

Chapter 6 Circularity

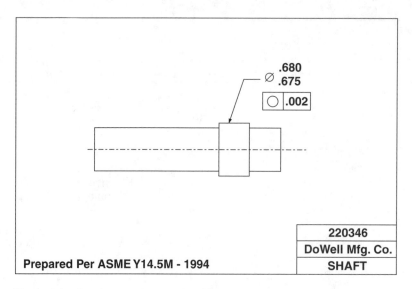

Figure 6.9 Drawing requirement for roundness on a shaft.

Figure 6.10 Measuring roundness on the journal shaft with a vee block.

VEE-ANVIL MICROMETER: OUTSIDE CIRCULARITY MEASUREMENT

Another effective circularity measurement method is to use a vee-anvil micrometer to measure the part. Although originally designed to measure three-fluted drill diameters, the vee-anvil micrometer can be used for effective circularity measurements as long as the discrimination of the micrometer does not exceed 10% of the circularity tolerance.

The same limitations apply to the vee-anvil micrometer as those mentioned in the vee-block method, but the vee-anvil micrometer is still superior to two-point instruments for measuring circularity. Most vee-anvil micrometers have a 60° angle, and so the same associated errors would be seen as with the vee-block method. The vee-anvil micrometer is most effective for measuring trilobed parts for circularity error (due to the 60° angle).

STEP 1 Measure the diameter of the part in several locations around a circular section, and memorize the difference in the actual measurements of that section, as shown in Figure 6.11.

STEP 2 Divide the difference by 3 to obtain the actual radial circularity error.

STEP 3 Repeat step 1 on other circular sections along the diameter. If the results are less than or equal to the circularity tolerance, the part is acceptable.

The vee-anvil micrometer is *not* recommended for parts that are even-lobed (for example, oval-shaped) for the same reasons as the vee-block method. Two-point micrometers will suffice for oval-shaped parts.

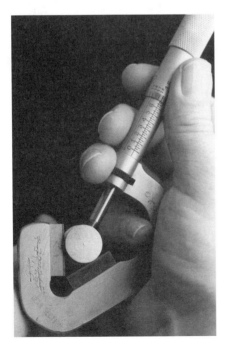

Figure 6.11 Measuring circularity on the journal shaft with a vee-anvil micrometer.

BORE GAGE METHOD: INSIDE CIRCULARITY MEASUREMENT

In the case of inside diameters, a variety of standard two-point instruments is used to measure the diameter (such as telescoping gages, small hole gages, and two-point bore gages), but they should *not* be used to measure circularity unless the inside diameter has an even number of lobes. Once again, a three-point measuring device is better for odd-lobed parts (especially trilobe parts). In these cases, instruments such as intramicrometers and three-point dial bore gages can be used for circularity measurements. The same limitations and correction factors previously covered apply, but circularity error is more readily detected with three-point instruments. A certain amount of error will result using three-point instruments on diameters that are not trilobed. In any event, the three-point device is better than any two-point device. The equipment necessary for this method is a three-point dial bore gage or an intramicrometer.

STEP 1 Insert the dial bore gage or intramicrometer into the hole at a given depth, and take a diametral measurement, as shown in Figure 6.12. Make at least three other measurements at the same depth (90° apart). Remember each actual reading. Divide the difference in diametral measurements made by the correction factor (or by 3 if necessary) to obtain the radial circularity error.

STEP 2 Repeat step 1 at different depths of the hole, continuing to the other end of the hole. If the difference in readings of the four directions is within the circularity tolerance, the part is acceptable.

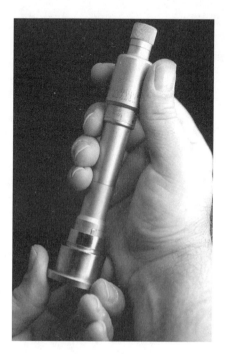

Figure 6.12 Circularity measurement on an inside diameter with a three-point intramicrometer.

VERIFYING CIRCULARITY USING RUNOUT TO CENTERS

Circularity tolerances can be *verified* using runout controls with respect to bench centers. (Refer to Chapters 11 and 12 for more information on runout measurement.) Circularity error affects runout error. To verify circularity means to provide assurance that the circularity tolerance has not been violated without direct measurement of circularity error. It is important to note that verifying circularity allows the observer to make an accept decision, but does not give the observer an actual circularity value (the reject decision cannot be made).

For example, if the circularity tolerance on a part diameter is .002″ (maximum is understood) and the diameter passes a .002″ maximum runout check on bench centers (as shown in Figure 6.13), the part is round with .002″. If the runout shown on the indicator is higher than .002″, it does not necessarily mean that the circularity is unacceptable; it may be a runout problem to the centers. (Refer to Chapter 11 on circular runout measurement for inspection methods.) Circularity can be verified (instead of measured) using runout when actual circularity values of the part are not required. Keep in mind that only the accept decision can be made in this case.

It is also important that some precautions be taken prior to verifying circularity on centers, such as the following:

1. Make sure that the centers themselves are aligned and free from wear (circularity problems within the center) and nicks.
2. Make sure the centers that have been drilled into the part are also aligned and free from damage such as nicks or raised material.

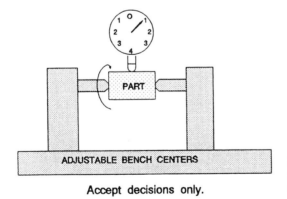

Accept decisions only.

Figure 6.13 Circularity is verified using runout measurements on bench centers.

ROTARY TABLE METHOD

A rotary table *(rotab)* can also be used to inspect circularity of inside or outside diameters since three points of contact are made by the rotab jaws. The rotab is a device that comes closer to the precision spindle method for measuring circularity. One important aspect when using a rotab for circularity measurement is to make sure that the axis of rotation of the rotab does not exceed 10% of the circularity tolerance.

When the axial error is equal or less than 10%, circularity can be accurately measured. Also, the error found by the indicator using a rotab does not require compensation because the part is fixed within the jaws of the rotab. The circularity can be measured directly. The equipment needed for this method is as follows:

1. Rotary table with axial error equal to or less than 10% of the circularity tolerance
2. Dial indicator that discriminates to 10% or less of the circularity tolerance
3. Surface plate
4. Precision gage pin (to check the rotab axial error)
5. Surface gage (or height gage) to hold the indicator

STEP 1 Check the axial error of the rotary table by placing a precision gage pin in the jaws and rotating the pin to obtain an FIM as shown in Figure 6.14. If the FIM of axial error is less than 10% of the circularity tolerance, proceed to step 2.

STEP 2 If the axial error of the rotab is acceptable (10% or less of part circularity tolerance), remove the gage pin and place the part in the rotab jaws.

STEP 3 Place the dial indicator on the part close to the jaws of the rotab and rotate the part to obtain an FIM, as shown in Figure 6.15. The FIM value obtained is the radial circularity error.

STEP 4 Repeat step 3 at other circular sections of the diameter. If the radial circularity error is equal to or less than the circularity tolerance, the part is acceptable.

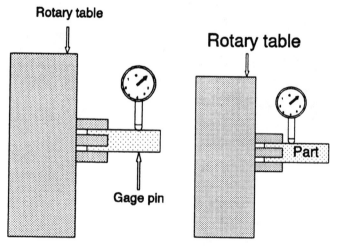

Figure 6.14 Checking the axis of the rotary table.

Figure 6.15 Measuring circularity using a rotary table.

The rotary table can also be effective for measuring the amount and location of lobes on the part in order to support the vee-block method previously discussed. As the part is being indicated in the jaws of the rotary table, indicator deviations (high and low) could be marked on the part that will help the observer draw a picture of lobe heights, quantities, and locations. As a final note, the effectiveness of the rotab circularity measurement can be improved by using a precision collet to hold the part and then locating the collet in the jaws of the rotab. In this manner, the contact made on the part is a full circle.

PNEUMATIC GAGES FOR CIRCULARITY MEASUREMENT

Circularity of inside diameters (bore diameters, for example) can be obtained by using pneumatic (air) gages with the proper number of air jets in the probe. Trilobed parts, for example, require three air jets in the probe. For five-lobed parts, five jets can be used for direct circularity measurements, or jets could be arranged in the probe such that the proper angle is generated (refer to the previous formula for selecting the proper vee-block angle); then the results can be converted using the same conversion equation that was used for the vee-block method of inspection.

Once the proper amount of jets is selected, master the air-gage probe on a master set ring; then insert the probe into the part at one circular element. Rotate the probe at least four times (90° apart) to find the highest and lowest readings. The results of the inspection should be divided by 3 to find the radial circularity error.

PRECISION SPINDLE METHOD

Precision spindles are, by far, the best instruments to measure circularity for many reasons:

1. Precision spindles are *radial* measurement instruments.
2. The actual number, height, and locations of lobes are known.
3. The precision spindle directly reproduces the two concentric circles of the tolerance zone and compares the circular sections of the part to the tolerance zone.
4. Errors in circularity measurement are at a minimum when precision spindles are used.
5. A graph of the resultant circularity error is produced.

Precision spindles are equipped with one or two probes that can be used for a variety of measurements. Some spindles have a probe on a precision vertical slide that can be used when making measurements of cylindricity, concentricity, and perpendicularity. There are two basic types of precision spindles: (1) the spindle rotates and the probe is stationary, and (2) the probe rotates and the spindle (base) is stationary. The two types of spindles are shown in Figure 6.16a and b.

Precision spindles rotate in a true circle within tight tolerances. For example, some air-bearing spindles can rotate in a true circle within .000005" (five-millionths). Various

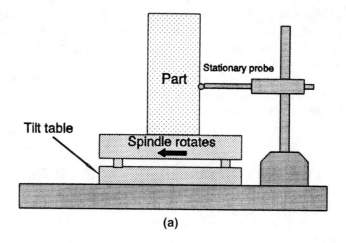

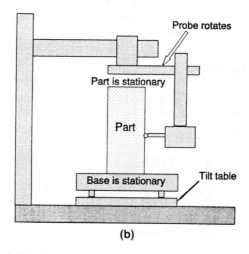

Figure 6.16 (a) A rotating-spindle machine, and (b) a rotating-probe machine.

makes and models of precision spindles can be purchased. Most have air-bearing spindles, are computer-assisted, and provide a printed polar graph of the results of each circular section measured. Precision spindles are very versatile and can also be used to inspect cylindricity, straightness, runout, concentricity, and other geometric controls. An example of a precision spindle is shown in Figure 6.17.

The setup for circularity measurement on most precision spindles involves the following general steps:

Figure 6.17 Precision spindle setup for measuring circularity. Courtesy of Federal Products Corp.

STEP 1 Set the magnification of the machine to the desired accuracy. The desired accuracy for the part is at least 10% of the circularity tolerance.

STEP 2 Insert a blank polar graph on the plotter of the machine when a polar graph is desired.

STEP 3 Choose the correct stylus (probe) for the application. For most applications, the stylus is 1/16" to 1/8" diameter. Refer to the operator's manual for each different instrument for further information. Place the part on the spindle and establish rough centering of the axis of the part to the axis of the spindle.

STEP 4 Bring the stylus in contact at a point on one circular element of the part and adjust the indication of the machine to zero. Many machines have automatic centering features.

STEP 5 Start the rotation of the machine and complete a full circle at the circular element. For machines that are computer-assisted, only rough centering will usually be necessary because the microprocessor on most machines provides automatic centering of the part.

STEP 6 After rough centering is accomplished, repeat steps 4 and 5 at other circular elements of the part. Figure 6.17 shows a precision spindle being used to measure circularity on a part diameter.

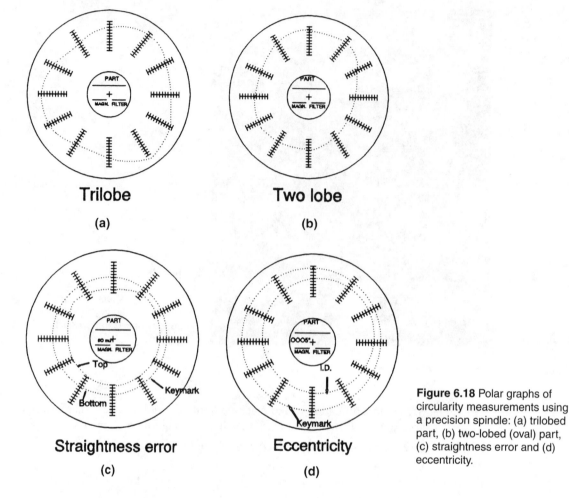

Figure 6.18 Polar graphs of circularity measurements using a precision spindle: (a) trilobed part, (b) two-lobed (oval) part, (c) straightness error and (d) eccentricity.

Polar Graphs

Although circularity error is shown on the digital readout of most machines, a polar graph is also optional in order to interpret the circularity in terms of lobes, and the like, or for a permanent record of the inspection. Figure 6.18 shows examples of actual polar graph results of precision spindle circularity measurements that have been cleaned and scanned for the purpose of clarity.

SUMMARY

Circularity inspection can be very involved when dealing with circularity tolerances in industry today and the effects (lobes) of the manufacturing process. It is not always feasible to purchase sophisticated equipment (for example, precision spindle equipment) for circularity inspection, nor, at times, is it necessary. On the other hand, it is necessary to avoid the use of simplistic measuring instruments (two-point instruments) for circularity measurement since they will surely be prone to considerable error.

REVIEW QUESTIONS

1. Using a two-point measuring device (for example, a micrometer or caliper) to measure a diameter that has odd-numbered lobes can result in significant size and circularity measurement error. *True* or *False?*
2. A polar graph can be obtained that shows circularity error using which of the following measurement equipment: (a) vee block, (b) precision spindle, (c) micrometer, (d) indexing head?
3. The imaginary tolerance zone for circularity (roundness) is two _____ circles the stated tolerance apart on their radii.
4. Circularity error is not additive to size; therefore, it cannot be used to accept size features that are oversize. *True* or *False?*
5. On which of the following features can circularity tolerances be applied if necessary in the design of the part: (a) inside diameters, (b) cones, (c) outside diameters, (d) any of the preceding?
6. Which of the following can cause the worst measurement error when using two-point measuring devices on odd-lobed diameters: (a) five-lobed part, (b) seven-lobed part, (c) three-lobed part, (d) two-lobed part?
7. When using a vee block for circularity measurement, other than the size of the vee block, what is another important characteristic of the vee block to consider?
   (a) The height
   (b) The parallelism of the faces
   (c) The location of the vertex of the angle
   (d) The angle of the vee
8. The resultant FIM reading, when using vee blocks for circularity inspection, can be used directly as a measurement of circularity error. *True* or *False?*
9. Pneumatic (air) gages can be used effectively for circularity measurement depending on the location of the _____ in the probe.
10. In the case of a part that has centers, circularity can be *verified* (not necessarily measured) using the _____ method.
11. Using only the following choices, which device is not capable of circularity measurement?
    (a) Precision spindle
    (b) Vee block and indicator
    (c) Plug gage
    (d) Rotary table
12. Rule 1 of the standard requires that circularity error must be contained within the _____ boundaries of the feature.
13. From the following methods, choose the best method for measuring circularity (roundness): (a) vee-block method; (b) vee-anvil micrometer; (c) rotary-table method; (d) precision spindle.
14. Which of the following lobe configurations allows accurate two-point circularity measurements: (a) trilobe, (b) five lobe, (c) two lobe (oval), (d) seven lobe?
15. Circularity tolerances apply only at each end of the diameter being measured. *True* or *False?*

7
Cylindricity

OBJECTIVES

Upon completion of this chapter, the reader should be able to:

1. Explain the difference between circularity and cylindricity.
2. Inspect cylindricity in an open setup (using methods from previous chapters also).
3. Verify cylindricity using total runout to centers.
4. Identify other geometric shapes that are inherently controlled by cylindricity.
5. Understand the application of a precision spindle and precision vertical slide for measuring cylindricity.

INTRODUCTION AND APPLICATIONS

Cylindricity is the most complex form tolerance of all and also the most difficult to inspect. Essentially, cylindricity is a combination of circularity at all circular elements, straightness of all surface elements, and taper, combined into one control. Cylindricity applications include all functional applications that require a true cylinder such as rotating shaft journal and bearing diameters, pistons and piston bores, and poppets. Cylindricity is often applied to bearing diameters of high-speed rotating parts, to qualify datum diameters, and to improve press or running fits.

In an open setup, cylindricity measurement is very time-consuming. Even with sophisticated equipment, a fair amount of complexity and time is required for best results. When cylindricity is required, we must take the available tools and do our best to inspect the product.

The drawing in Figure 7.1 shows a cylindricity requirement. The feature control frame states that the diameter must be a true cylinder within a .002″ radial tolerance zone. The tolerance zone for this cylindricity requirement is *two concentric cylinders that are .002″ apart on their radius,* as shown in Figure 7.2.

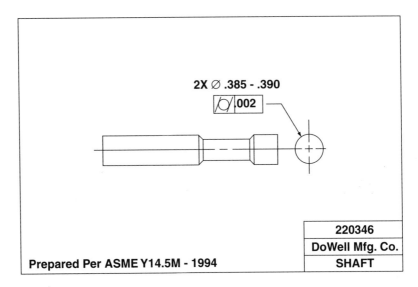

Figure 7.1 Cylindricity requirement on a drawing.

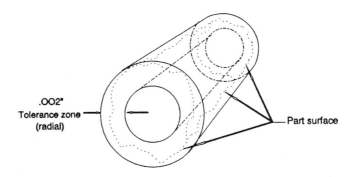

All surface elements of the part diameter must lie within the annular space between the two cylinders.

Figure 7.2 Tolerance zone for cylindricity is two concentric cylinders.

KEY FACTS

1. Cylindricity tolerance is not associated with a datum reference.
2. Cylindricity is a radial tolerance zone defined by two perfectly concentric cylinders the stated tolerance apart on their radii.
3. Cylindricity automatically controls circularity at each circular element, straightness of surface elements, barrel-shaped or hourglass-shaped conditions, and taper due to the tolerance zone of two concentric cylinders.
4. Cylindricity tolerances cannot violate the maximum boundary of perfect form at MMC of the associated size tolerance.

OPEN-SETUP METHODS

For practical purposes, if the observer were to inspect *circularity at several circular elements, straightness of surface elements at several line elements, and taper*, the cylindricity of the part could be inspected. (Refer to Chapters 5 and 6 for straightness and circularity measurement methods, respectively.) Also note that barrel shapes and hourglass shapes are controlled by cylindricity tolerance. During the exploring measurements for taper, the observer should also look for these conditions.

Since these methods have already been covered in Chapters 5 and 6, further discussion is not required. In an open setup, measurements of surface element straightness, circularity at several circular elements, and taper per side can be used to estimate the cylindricity of the part. For example, if a part has a cylindricity tolerance of .002″, the tolerance zone is two concentric cylinders that are .002″ apart on their radius. If the part is inspected in open setup and found to be round within .002″ (at the worst element), straight on its surface elements within .002″, and not tapered more than .002″ *per side*, the part is cylindrical within .002″.

CYLINDRICITY VERIFIED USING TOTAL RUNOUT

When open-setup measurements are required and there is no need for an actual value of cylindricity, total runout with respect to bench centers is a better open-setup method. Total runout, when applied to a part diameter, controls many different geometric errors, including concentricity, circularity, taper, cylindricity, and straightness of surface elements.

For this reason, total runout can be used to verify if cylindricity is within specifications, because if the composite of all these geometrical errors does not exceed the cylindricity requirement, then cylindricity must be within specifications. It must be understood that this method verifies only that cylindricity is within tolerances; it does not measure cylindricity or allow the observer to reject cylindricity. Only the acceptance decision can be made. Once again, when actual cylindricity values are not necessary, verifying cylindricity using total runout is a fast and efficient method. The part, of course, must have centers to use this method.

The equipment needed for this method is as follows:

1. Surface plate
2. Bench centers
3. Indicator that discriminates to no more than 10% of the cylindricity tolerance
4. Height gage (or surface gage) to hold the indicator
5. Precision parallel (or similar device) to guide the base of the height gage or indicator in a straight line with respect to the centers

STEP 1   Check the bench centers and the part, and make sure the centers and the drilled centers in the part are free from damage or dirt.

STEP 2   Secure the part between the bench centers.

STEP 3   Place the indicator tip at top dead center on one end of the part and set zero on the indicator. Make sure the height gage (or surface gage) base is referenced against the centers (or a precision parallel) in a manner such that the indicator can be guided in a straight line parallel to the axis of the centers, as shown in Figure 7.3.

STEP 4   While rotating the part, traverse the indicator across the entire length of the part in a straight line (guided) and look for a full indicator movement (FIM). If the FIM is less than or equal to the cylindricity tolerance, the part is within cylindricity tolerance. (*Note:* If the FIM exceeds the cylindricity tolerance, the part cannot be rejected using this method. The excessive FIM may be the result of poor runout, poor eccentricity of the centers, or other geometrical errors not caused by cylindricity problems. Only the acceptance decision, that is, FIM is less than or equal to the tolerance, can be made using this method. For more information on total runout, refer to Chapter 12.)

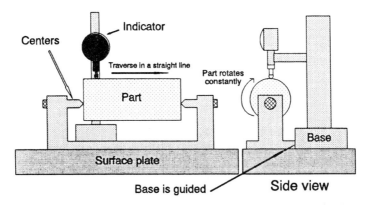

Figure 7.3   Verifying cylindricity using total runout on centers.

PRECISION SPINDLE METHODS

The best method for measuring cylindricity is a precision spindle *that is also equipped with a precision vertical slide.* The typical precision spindle affords excellent circularity inspection, but it is still difficult to relate all the circularity plots to each other with respect to the two imaginary concentric cylinders of tolerance (even with the polar graphs provided by the machine). Multiple polar plots would be necessary on a standard precision spindle, and we would have to relate them to each other for cylindricity measurement.

If the spindle had a precision vertical slide (as shown in Figure 7.4), the probe (or stylus) could maintain constant contact with the part as the slide sweeps vertically. Therefore, a constant trace of the part could result on the polar graph.

The use of a precision spindle for cylindricity measurement must be based on the type of spindle used, the accessories that come with the spindle, and the ability of the spindle's computer (if the spindle has one) to relate polar graphs to each other for cylindricity evaluation. Refer to the manufacturer's operations manual for the spindle for best results. For open-setup methods, refer to the chapters cited previously. Cylindricity can be measured if enough care is taken to relate the various geometric errors of the part to the two concentric cylinders of tolerance.

In Figure 7.4, the probe of the precision spindle has been set at the bottom of the diameter of the part in order to begin the process of measuring the cylindricity of the

Figure 7.4 A precision spindle with a precision vertical slide for cylindricity. Courtesy of Federal Products Corp.

Precision Spindle Methods 101

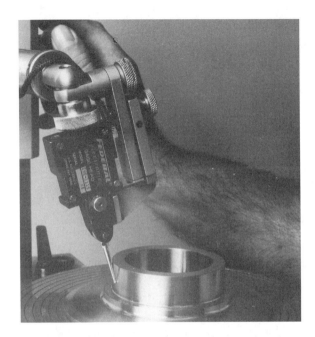

Figure 7.5 The probe is at midposition on the diameter. Courtesy of Federal Products Corp.

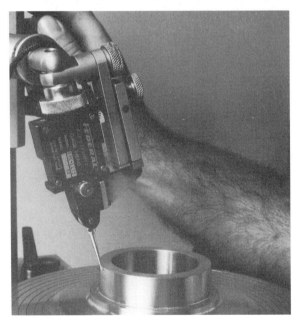

Figure 7.6 The probe is at final position on the diameter. Courtesy of Federal Products Corp.

part. Figures 7.5 and 7.6 show some progressive positions of the probe during cylindricity measurement. While the spindle is rotating, the vertical slide allows probing of the entire diameter. Using the computer and polar graph, the observer can obtain graphs of the cylindricity of the part.

REVIEW QUESTIONS

1. The tolerance zone for cylindricity is two concentric _____ the stated tolerance apart on their radii.
2. Which of the following is controlled automatically by cylindricity tolerance when it is applied to a cylindrical size feature: (a) location, (b) parallelism, (c) taper, (d) size?
3. Cylindricity can be "verified" (not directly measured) by using total _____.
4. Cylindricity, due to its inherent controls, cannot be applied to (a) inside diameters, (b) outside diameters, (c) cones, (d) holes.
5. One of the best methods for accurate measurement of cylindricity requirements is a (a) vee block and indicator (or probe), (b) ring gage, (c) plug gage, (d) precision spindle.
6. Which of the following are inherently controlled by cylindricity tolerance?
   (a) Circularity, straightness, and taper
   (b) Circularity, location, and size
   (c) Size, taper, and location
   (d) Circularity, parallelism, and size
7. It would be proper for a designer to apply cylindricity to the inside diameter of a flat washer. *True* or *False?*
8. Cylindricity is not associated with a datum reference. *True* or *False?*
9. Cylindricity controls both the straightness of surface elements and the straightness of the axis of a diameter. *True* or *False?*
10. Using a precision spindle for cylindricity is more effective when the spindle is accompanied by a precision _____ slide.

8

Parallelism

OBJECTIVES

Upon completion of this chapter, the reader should be able to:

1. Explain the various parallelism tolerance zones for different applications.
2. Explain how parallelism of a surface inherently controls the flatness of that surface.
3. Explain the difference between bonus tolerances on controlled features and additional tolerances from size datums.
4. Explain how parallelism of a surface is not additive to associated size tolerances.
5. Measure surface-to-surface parallelism on a surface plate.
6. Measure axis-to-surface parallelism on a surface plate.
7. Understand functional gaging of parallelism of size features at maximum material condition (MMC).

INTRODUCTION AND APPLICATIONS

Parallelism is an *orientation tolerance* that relates a surface, centerplane, or an axis parallel to a datum plane or axis. Parallelism is often used as a refinement of size tolerances on noncylindrical parts for a variety of fit, form, or functional reasons, such as control of the stackup of tolerances or of two interrelated surfaces. Parallelism control requires at least one primary datum reference (at times, more than one datum is necessary).

The tolerance zone for parallelism depends largely on the feature being controlled *and* the type of datum. For example, if the controlled feature is a surface and the datum feature is a surface, the tolerance zone is *two imaginary parallel planes the stated tolerance apart*. Another example is parallelism of the axis of a hole to the axis of another hole. In this case, the imaginary tolerance zone is a *cylinder*. Contrary to popular belief, the tolerance zone for parallelism of an axis is *not always a cylinder*. If the datum reference for the parallelism of an axis is a plane (such as a surface), the tolerance zone for the parallelism of the axis is still *two parallel planes the stated tolerance apart*. Figure 8.1 shows these three variations of parallelism tolerance zones. One fast method of remembering the different parallelism tolerance zones is as follows:

1. If the datum is *a plane*, the parallelism tolerance zone is *two planes*.
2. If the datum is *an axis*, the parallelism tolerance zone is *a cylinder*.

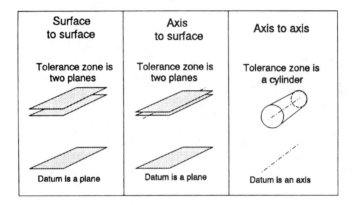

Figure 8.1 Parallelism tolerance zones.

KEY FACTS

1. Parallelism tolerance must be called out to at least one datum (sometimes more than one).
2. The tolerance zone for parallelism depends heavily on the type of datum and the controlled feature as covered in the introduction (see Figure 8.1).
3. Parallelism (surface to surface) cannot be gaged; it must be measured. Parallelism of size features at MMC can be gaged with a functional gage.
4. In most cases, the results of parallelism error are stated in terms of full indicator movement (FIM).
5. A parallelism tolerance applied to a surface inherently controls the flatness of that surface to the extent of the stated parallelism tolerance.
6. Parallelism tolerance cannot violate the maximum boundary of perfect form at MMC for the related size tolerance.
7. Parallelism inspection is accomplished by mounting the datum on a simulated plane and then indicating the controlled surface (FIM).

SURFACE-TO-SURFACE PARALLELISM: SURFACE-PLATE METHOD

The most straightforward of all parallelism measurements is plane-to-plane parallelism. In the past, parallelism has been misunderstood by some observers to the extent that various thickness measurements were made in an effort to inspect parallelism. Parallelism is measured with respect to a datum plane that must be properly contacted in order to be a functional measurement. A variety of thickness measurements with a micrometer or caliper cannot establish the highest points of the functional (primary) datum. Parallelism is still, however, a fairly simple measurement.

Refer to the part shown in Figure 8.2. The feature control frame states that the controlled surface must be parallel within .002" with respect to primary datum A. The pri-

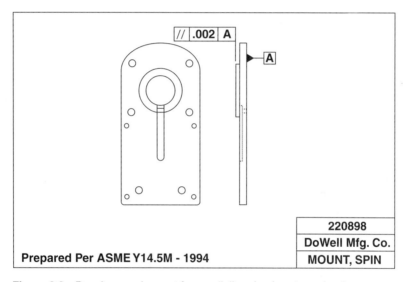

Figure 8.2 Drawing requirement for parallelism (surface to surface).

mary datum must be contacted by at least the three highest points on the datum feature (A). The tolerance zone for this relationship consists of *two perfectly flat and parallel planes that are .002" apart*, as shown in Figure 8.3.

The following is a list of equipment necessary for this measurement:

1. Surface plate (or bench comparator surface)
2. Dial indicator that discriminates to no more than 10% of the parallelism tolerance
3. Surface gage (or height gage) to hold the indicator

STEP 1 Rotate and slide the part and equipment carefully onto the surface plate so that the datum feature is resting on the plate. Refer to Figure 8.4 for proper methods of putting parts and gages onto and removing them from the surface plate (in order to avoid damaging the part or the surface plate).

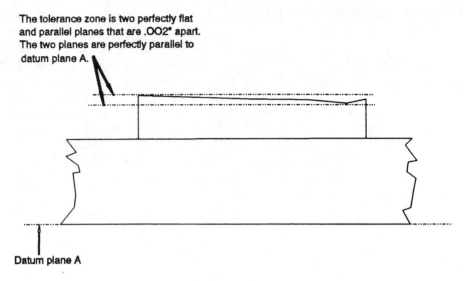

Figure 8.3 Tolerance zone for parallelism.

Figure 8.4 Proper method of (a) putting parts or gages on a surface plate (rock and slide them on).

Surface-to-Surface Parallelism: Surface-Plate Method

Figure 8.4 (*continued*) (b) Proper method of removing parts and gages from the surface plate (slide them off).

STEP 2 Set the tip of the dial indicator on the controlled surface; set zero on the indicator and then sweep the entire surface looking for the FIM, as shown in Figure 8.5. (*Note:* Setting zero on the indicator is not really necessary, but

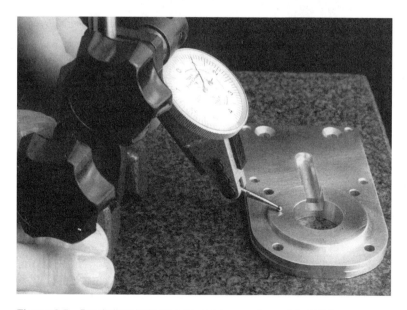

Figure 8.5 Parallelism being measured on a surface plate (FIM).

108   Chapter 8   Parallelism

it establishes the travel necessary for the FIM measurement, and it is a good starting point.) The FIM of the surface should not exceed the parallelism tolerance. If the FIM found is less than or equal to the parallelism tolerance, the part is acceptable.

Simple surface-to-surface parallelism is nothing more than establishing at least the three highest points of the primary datum surface and then sweeping the controlled surface with an appropriate indicator looking for the FIM of the surface to the datum plane. The surface plate establishes the datum by finding automatically the three or more highest points of the datum feature (A in this case). Another example of parallelism measurement is shown in Figure 8.6.

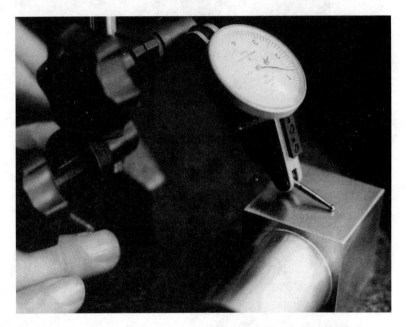

Figure 8.6 Another part is measured for surface-to-surface parallelism.

AXIS-TO-SURFACE PARALLELISM: SURFACE-PLATE METHOD

For many functional reasons, design engineers may require that the axis of a hole (or a pin) be parallel to a datum surface. The control, in this case, has nothing to do with the location tolerance of the axis of the hole (that might be accomplished using position tolerance). The axis of the hole in this case is an orientation tolerance that ensures that the hole is aligned with a datum plane. Once again, parallelism controls only the orientation of the hole (or pin) to the datum plane, not the location. (Refer to the drawing in Figure 8.7.) The feature control frame states that the axis of the hole must be parallel to the datum surface within .005″ (FIM is understood). The datum is a plane, so the

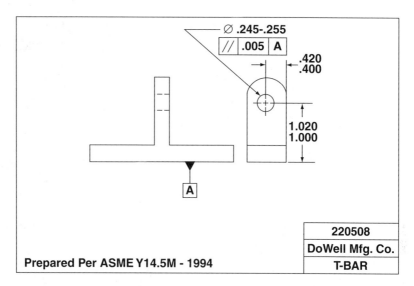

Figure 8.7 Drawing requirement for parallelism of an axis of a hole to a datum plane.

imaginary tolerance zone for the axis is *two imaginary and perfectly parallel planes that are .005" apart. These planes are perfectly parallel to the datum plane.* Since this requirement is specified regardless of feature size (RFS), functional gaging cannot be used; the part must be measured for parallelism.

In this case, an open-setup measurement for the part consists of the following necessary equipment:

1. Surface plate
2. Dial indicator that discriminates to no more than 10% of the parallelism tolerance
3. Surface gage (or height gage) to hold the indicator
4. Largest gage pin that will fit into the controlled hole

STEP 1 Select the largest gage pin that will fit the controlled hole, and insert it carefully into the hole. Slide the part carefully onto the surface plate.

STEP 2 Using the dial indicator and surface gage (or height gage), find top dead center of the pin immediately next to one entrance of the hole and set zero, as shown in Figure 8.8. Then move the indicator over to the opposite entrance of the hole and take an FIM reading at top dead center, as shown in Figure 8.9. The FIM value must be equal to or less than the .005" tolerance for parallelism.

Since the tolerance zone for this application is two planes with respect to a datum plane, infinite rotation is allowed, and the parallelism tolerance need not be measured from side to side, or in any other direction. The gage pin was used to make sure that the functional axis of the hole was being measured, thereby assuring that the measurement made was functional and represented the use of the part in assembly.

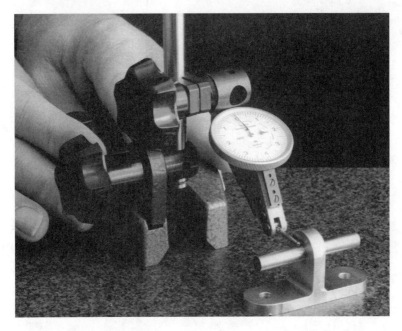

Figure 8.8 Setting zero at one end of the pin (at top dead center).

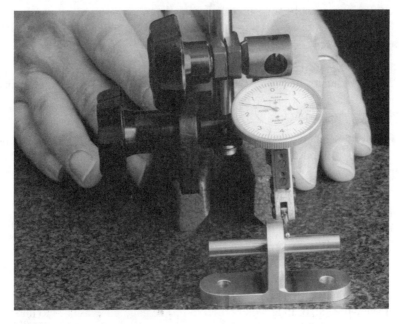

Figure 8.9 Finding the FIM at the opposite entrance of the hole (at top dead center).

AXIS-TO-SURFACE PARALLELISM: BONUS TOLERANCES

The drawing of the part previously used could be altered to reflect parallelism of the hole within .005″ at MMC (as shown in Figure 8.10), thereby allowing for the option of using a functional gage, or we could apply bonus parallelism tolerance based on the actual size of the controlled hole. The tolerance zone description is *two imaginary and perfectly parallel planes that are .005″ apart (when the hole is at MMC size)*, as shown in Figure 8.11.

Since MMC modifies the tolerance zone, the option here is to measure the parallelism as previously discussed and allow for the bonus tolerance or to design a functional gage. If the part were measured on the surface plate, the table in Figure 8.12 could be used to allow for bonus tolerances. In this case, it is very important that the functional gage be designed in a manner such that the virtual size pin for the controlled hole is floating (remember, position is not being controlled). The only extra work in open setup is to measure the actual size of the hole to determine the bonus tolerance that applies.

If the part is within the stated parallelism tolerance zone, the bonus tolerance is not needed and therefore not required to be calculated. If the part is outside the stated tolerance, the hole size can be measured, and the bonus tolerance (if any) can be added in order to make the acceptance decision. The bonus parallelism tolerance (due to hole size) that can be added to the stated parallelism tolerance is equal to the amount that the actual size of the hole departs from the MMC size of the hole.

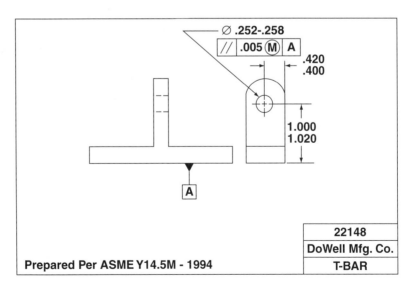

Figure 8.10 Parallelism requirement at MMC. Bonus tolerances are applicable if needed.

112   Chapter 8   Parallelism

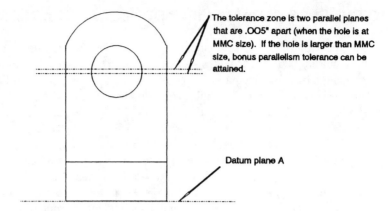

Figure 8.11   Tolerance zone for the part.

Actual Hole Size (")	Bonus Tolerance Allowed (")	Total Parallelism Tolerance Allowed (")
.252	.000	.005
.253	.001	.006
.254	.002	.007
.255	.003	.008
.256	.004	.009
.257	.005	.010
.258	.006	.011

Figure 8.12   Table of bonus tolerances for the part parallelism tolerance.

For example, in this case, if the hole size is larger than its MMC limit by .003", the bonus parallelism tolerance is .003". The .003" can then be added to the stated .005" parallelism tolerance for a total of .008" parallelism tolerance. (*Note:* The hole cannot be oversize. Bonus tolerance amounts can only use the hole size tolerance range.)

AXIS-TO-AXIS PARALLELISM: SURFACE-PLATE METHOD

Another example of parallelism is when the axis of a size feature must be parallel to the axis of a datum size feature. For example, the axis of a hole must be parallel to the axis of another hole on the part. An example of this application is shown in Figure 8.13.

Since the parallelism is called out regardless of feature size (RFS)—(RFS is understood per rule 2)—the part must be measured. It cannot be gaged with a functional

Axis-to-Axis Parallelism: Surface-Plate Method 113

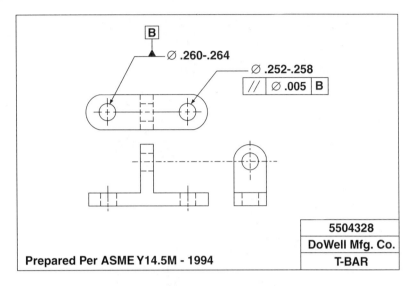

Figure 8.13 Drawing requirement for parallelism of an axis to an axis at RFS.

gage. It is important to note that, since the datum for parallelism is an axis, the tolerance zone for control of parallelism of the controlled axis is a *cylinder*.

In this case, the parallelism can be measured using the following equipment:

1. Surface plate
2. Dial indicator that discriminates to no more than 10% of the parallelism tolerance
3. Surface gage (or height gage) to hold the indicator
4. Set of matched vee blocks (or one vee block with a clamp)
5. Two gage pins; one is the largest gage pin that will fit into the datum hole, and the other is the largest gage pin that will fit into the controlled hole

STEP 1 Insert the gage pins selected into the controlled hole and the datum hole.

STEP 2 Slide the surface gage and the vee blocks onto the surface plate.

STEP 3 Mount the datum pin onto the matched vee blocks, as shown in Figure 8.14, making sure that the controlled hole is accessible for measurement.

STEP 4 Indicate the parallelism of the controlled hole at both the entrance and the exit (top dead center), as shown in Figure 8.15. The maximum parallelism error allowed is equal to the stated parallelism tolerance.

STEP 5 Rotate the setup horizontally, and indicate the parallelism (top dead center) once again, as shown in Figure 8.16. The maximum parallelism error allowed is equal to the stated tolerance.

STEP 6 If the parallelism error in all directions is within the stated parallelism tolerance, the part is acceptable.

Figure 8.14 Datum pin and part are mounted in the vee block.

(a)

Figure 8.15 (a) and (b) Parallelism of the hole being checked in the vertical position (at each side of the hole).

Figure 8.15 *(continued)*

Figure 8.16 (a) and (b) Parallelism of the hole being checked in the horizontal position (at each side of the hole).

(b)

Figure 8.16 *(continued)*

PARALLELISM: SECONDARY ALIGNMENT DATUMS

Parallelism, at times, is called out with respect to more than one datum, especially when more than one surface is functionally involved in assembly. In such cases, the parallelism of a surface is still a relatively straightforward inspection, except that the primary and secondary datum must first be properly contacted. An example of parallelism using a secondary alignment datum is shown in Figure 8.17.

The part in Figure 8.17 is a parallelism callout; the primary datum must first be contacted at a minimum of the three highest points on the surface and then the secondary datum must be contacted at the two highest points. The equipment necessary for this inspection is as follows:

1. Surface plate
2. Dial indicator that discriminates to no more than 10% of the parallelism tolerance
3. Height gage (or surface gage) to hold the indicator
4. Right-angle plate (knee) or precision parallel (or similar surface-plate accessory) that will make full contact with the secondary alignment datum surface

STEP 1 Slide the part, knee, and surface gage carefully onto the surface plate, making sure that the primary datum feature on the part is resting on the surface plate.

Parallelism: Secondary Alignment Datums   **117**

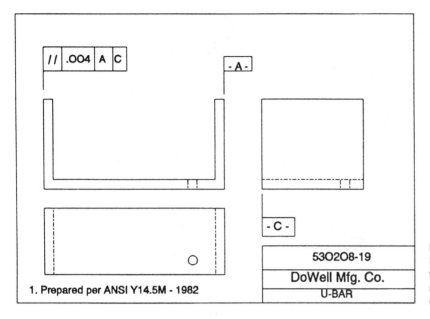

Figure 8.17 Parallelism requirement that specifies a secondary alignment datum.

STEP 2   Bring the secondary datum of the part in contact with the knee (or precision parallel), as shown in Figure 8.18.

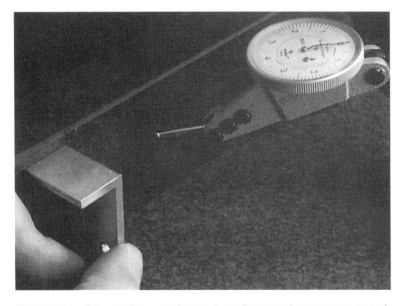

Figure 8.18   Primary datum *and* secondary alignment datum are contacted.

118 Chapter 8 Parallelism

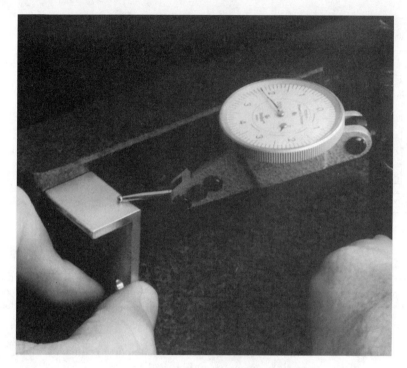

Figure 8.19 Sweeping the surface for parallelism error (FIM).

STEP 3   Sweep the indicator across the entire controlled surface to obtain an FIM of parallelism error, as shown in Figure 8.19. If the part is equal to or less than the parallelism tolerance, it is acceptable.

The primary difference in simple parallelism applications and secondary alignment datum applications is the contact of the secondary datum prior to inspection.

PARALLELISM: COMPOUND DATUMS

At times, the parallelism of a surface must be referenced to an interrupted surface (see Figure 8.20) or two offset surfaces (see Figure 8.21). In these cases, the primary datum is still defined as the three highest points, except that they are the three highest points when contacting two different surfaces. For example, two high points may be on one of the surfaces and the third on another (and vice versa). Notice that the drawing with *offset compound datums* shows a basic dimension that defines the offset.

The equipment needed to measure the part in Figure 8.20 is the following:

1. Surface plate
2. Dial indicator that discriminates to no more than 10% of the parallelism tolerance
3. Surface gage (or height gage) to hold the indicator

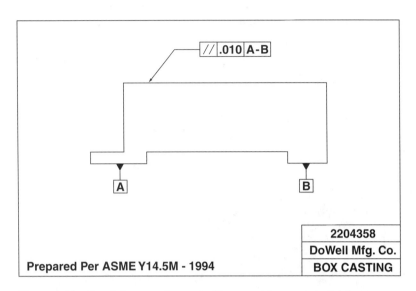

Figure 8.20 Parallelism requirement with respect to compound datums.

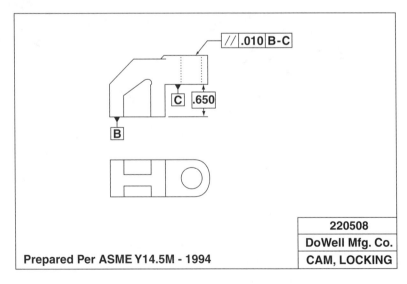

Figure 8.21 Parallelism requirement with respect to offset compound datums. (*Note:* In manufacturing, the .650″ basic dimension must be held within tooling tolerances).

120 Chapter 8 Parallelism

STEP 1 Locate the part on the compound datums on the surface plate, as shown in Figure 8.22

STEP 2 Sweep the controlled surface with the dial indicator, and if the FIM is no larger than the stated tolerance, the part is acceptable (see Figure 8.23).

Figure 8.22 Primary (compound) datum is established on a surface plate.

Figure 8.23 Parallelism is being inspected (FIM) to the compound datums contacted.

For compound datums that are offset (as shown in the drawing in Figure 8.21), the offset has to be established with another plane at the basic offset dimension. The equipment needed to measure the part in Figure 8.21 is the following:

1. Surface plate
2. Dial indicator that discriminates to no more than 10% of the parallelism tolerance

Parallelism: Compound Datums   **121**

3. Surface gage (or height gage) to hold the indicator
4. Gage block stack of the basic dimension on the drawing that defines the offset

STEP 1   Wring a gage block stack to the height of the basic offset dimension identified on the drawing. (*Note:* For other larger parts, a precision parallel could rest on two equal gage block stacks in order to establish the basic offset dimension between the offset datums.) Take care to account for the thickness of the precision parallel when selecting the gage blocks.

STEP 2   Locate the compound datums for inspection with one datum on the surface plate and the other on the gage block stack, as shown in Figure 8.24.

STEP 3   Sweep the controlled surface and watch for the FIM, as shown in Figure 8.25. If the FIM is no more than the stated parallelism tolerance, the part is acceptable.

Figure 8.24   Compound offset datums are located. The part is ready for inspection.

Figure 8.25   Parallelism being inspected to compound offset datums.

In summary, parallelism is one of the easiest of geometric tolerances to inspect due to the simplicity of the requirement (FIM values) and to the fact that datums are always used in the relationship. As noted earlier, the most complex parallelism requirements are those that require secondary alignment datums, but they have not been specified on the drawing.

REVIEW QUESTIONS

1. The imaginary tolerance zone for parallelism depends largely on whether the datum is a plane or an axis. *True* or *False?*
2. What is the tolerance zone for parallelism of the axis of a hole to a datum plane?
   (a) Two parallel planes
   (b) A cylindrical zone
   (c) Two parallel lines
   (d) A circle
3. Parallelism tolerances always require a datum reference. *True* or *False?*
4. Parallelism tolerances are part of the form tolerance family. *True* or *False?*
5. What other geometric tolerance is automatically controlled when parallelism is applied to a surface?
   (a) Position
   (b) Symmetry
   (c) Flatness
   (d) Profile
6. The resultant measurement of parallelism is a(n) _____ value.
7. Surface-to-surface parallelism can be measured by taking multiple thickness measurements on the two surfaces with a micrometer. *True* or *False?*
8. Some parallelism tolerance measurements can be in error due to misalignment because of the lack of a _____ alignment datum.
9. When parallelism is applied to size features at MMC, a bonus tolerance can be allowed as the size feature departs from its MMC size. *True* or *False?*
10. Parallelism of the axis of a hole to a surface also controls the location of the hole. *True* or *False?*
11. Which of the following methods can be used to inspect the parallelism of a hole to a datum hole specified at MMC?
    (a) Open setup measurements
    (b) Functional gage
    (c) Neither a nor b
    (d) Both a and b
12. If a surface has been identified as the datum for the parallelism of another feature, that datum must be contacted at three or more of the _____ points on the surface.
13. Parallelism error cannot be used to violate the maximum _____ of perfect form when size tolerances are involved (reference general rule 1).
14. Surface-to-surface parallelism cannot be gaged; it must be measured (FIM). *True* or *False?*
15. It is possible for a parallelism requirement to have more than one datum specified (even though many applications of parallelism only use one datum). *True* or *False?*

9

Perpendicularity

OBJECTIVES

Upon completion of this chapter, the reader should be able to:

1. Explain the tolerance zone for perpendicularity for a given surface or axial application.
2. Perform the following perpendicularity inspection methods: precision square, cylindrical square, and right-angle plate.
3. Understand functional gaging of size features that must be perpendicular at MMC and the associated virtual condition.
4. Explain, for perpendicularity of plane surfaces, the importance of a secondary alignment datum.
5. Explain how vertical measurement systems work to measure perpendicularity.

INTRODUCTION AND APPLICATIONS

Perpendicularity applies to features of products that have a 90° relationship for functional reasons such as a dual interface in assembly, mating parts with holes for fasteners that share a common interface, refining a position tolerance, and many other applications involving two related features.

Perpendicularity is also used to control the relationship between two datum features (therefore *qualifying* them) to be used as datums (for example, primary and secondary). Perpendicularity (often referred to as *squareness*) is an orientation tolerance, and at least one datum reference must be shown for perpendicularity tolerances (at times, two datums are needed). Surface-to-surface perpendicularity measurement must verify that the controlled surface lies between *two parallel planes the stated tolerance apart and the planes of tolerance are exactly 90° from the datum plane*.

KEY FACTS

1. Perpendicularity tolerances must be related to a specific datum reference (in some cases, more than one).
2. Perpendicularity applies to features that must be 90° apart.
3. For controlled surfaces or centerplanes, the perpendicularity tolerance zone is two parallel planes the stated tolerance apart.
4. For controlled axes, the perpendicularity tolerance zone is two parallel planes of infinite rotation and is usually clarified in the feature control frame as a cylinder.
5. Perpendicularity of size features at maximum material condition (MMC) can be gaged, but other applications—for example, planes or regardless of feature size (RFS features) must be measured.
6. A perpendicularity tolerance applied to a surface controls the flatness of that surface to the extent of the stated perpendicularity tolerance.

IMPLIED PERPENDICULARITY

Fundamental rules of the standard require specific interpretation of 90° angles, as shown on the engineering drawing. Two rules apply to 90° (right) angles; both rules indicate that for all features that are shown at right angles, a 90° angle is understood and need not be specified. The first rule states that "*a 90° angle applies where center lines and lines depicting features are shown at right angles, and no angle is specified.*" This means that the designer does not have to specify 90° angles and that these angles are interpreted as nominal 90° angles where general tolerances for angles apply. The other rule covering right angles states that "*a 90° basic angle applies where center lines of features in a pattern or surfaces, shown at right angles on a drawing, are located or defined by basic dimensions and no angle is specified.*"

Figure 9.1 presents several 90° angles, one of them with a perpendicularity requirement. The drawing also has a title block table that shows tolerances for angles (± 2°) that apply (unless otherwise specified). Following the two implied 90° rules, all of the 90° angles are interpreted as 90° plus or minus 2° except for the surfaces involved with the perpendicularity requirement. In that case, the 90° angle is basic, and the tolerance is .005″.

PERPENDICULARITY: SINGLE SURFACE, ONE DATUM

The first example part for perpendicularity measurement is shown in Figure 9.1. The feature control frame states that the controlled surface must be perpendicular within .005″ to datum plane A. Therefore, the entire controlled surface must lie within a tolerance zone of *two parallel planes that are .005″ apart and perfectly perpendicular to datum plane A*, as shown in Figure 9.2. Several methods can be used to inspect this part (as it cannot be functionally gaged). In each case, however, the method must prove that the controlled surface is within the imaginary tolerance zone—most methods use full indicator movement (FIM) measurements.

Perpendicularity: Single Surface, One Datum   **125**

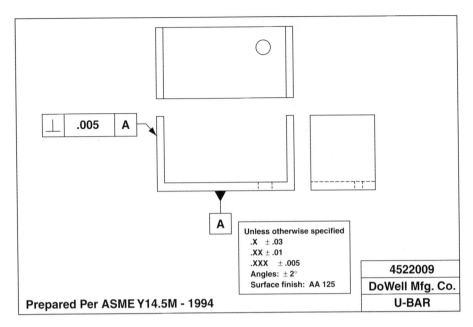

Figure 9.1 Drawing requirement for surface-to-surface perpendicularity.

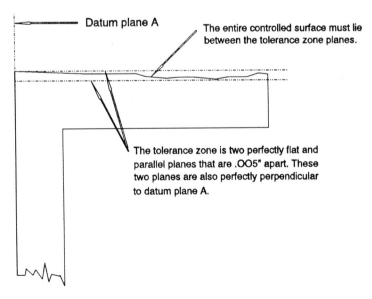

Figure 9.2 Tolerance zone for the part.

Figure 9.3 The part and precision square are located on the surface plate.

PERPENDICULARITY: PRECISION-SQUARE METHOD

A precision square (shown in Figure 9.3) is a simple method of inspecting perpendicularity (when the tolerance is relatively loose). The precision square has a base and a vertical blade at 90° that is perpendicular within gaging tolerances. The square is usually used in conjunction with precision feeler stock (preferably feeler wire) to make the comparison. The equipment required for inspection is as follows:

1. Surface plate
2. Precision feeler stock (preferably feeler wire in increments that discriminate to no more than 10% of the perpendicularity tolerance)
3. Precision square

This commonly used method is very straightforward, as seen in the following steps:

STEP 1 Locate the part primary datum and the base of the precision square on the surface plate, as shown in Figure 9.3.

STEP 2 Bring the blade of the precision square against one vertical line element of the controlled part surface, and probe the entire length with various thicknesses of feeler wire looking for gaps (see Figure 9.4).

STEP 3 Repeat step 2 on several other line elements of the controlled surface.

PERPENDICULARITY: CYLINDRICAL-SQUARE METHOD

A more sophisticated direct method of measurement of surface-to-surface perpendicularity involves the use of a precision cylindrical square instead of a precision square. A

Figure 9.4 Inspecting perpendicularity with a precision square and feeler stock.

typical cylindrical square has a diameter of 5″ and a height of 12″. A direct-reading cylindrical square (see Figure 9.5) is constructed in such a manner that one end of the square is flat and perpendicular to the axis of the square, but the other end is flat and at a small known angle to the axis.

When the square is placed on the end machined at an angle, it can be used to read perpendicularity error directly within .0002″ over the 12″ length. Of course, since

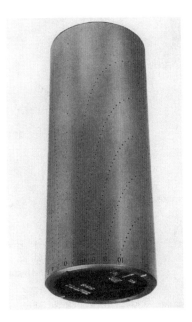

Figure 9.5 Precision cylindrical square. Courtesy of Brown and Sharpe Mfg. Co.

cylindrical squares are considerably more expensive than precision squares, this factor must also be taken into account.

The only equipment needed for this measurement is a surface plate and a cylindrical square. The method of inspection is as follows:

STEP 1   Mount the part datum on the surface plate.

STEP 2   Mount the cylindrical square on the surface plate on the end that is not perpendicular to the axis of the square.

STEP 3   Move the edge of the cylindrical square to make contact with the controlled surface of the part (as shown in Figure 9.6a); then rotate the square to find the error position on the square that matches (or is parallel to) the error of the vertical line element of the part (all light between the edge of the square and the surface of the part is eliminated, as shown in Figure 9.6b). Feeler stock is not required. Once the edges of the part surface and the square have been matched (no light), the highest point of the part surface will be in proximity with one of the out-of-squareness dotted lines on the cylindrical square. The observer can, at this time, read the perpendicularity error of the part directly from the square.

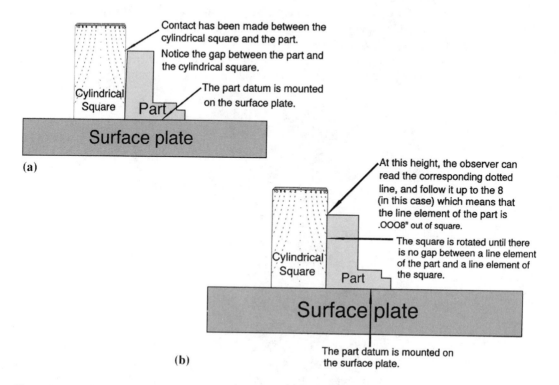

Figure 9.6   (a) Setup for using a cylindrical square, and (b) the cylindrical square and part surface are matched.

STEP 4 Since the surface of the part is being controlled, the observer will have to repeat step 3 at several line elements of the controlled surface. If each line element measured is at or below the perpendicularity tolerance, the part is acceptable.

SPECIAL INDICATOR STANDS FOR PERPENDICULARITY INSPECTION

Another method for using the cylindrical square is to use the end of the square that is perpendicular to the axis as a master to set an indicator mounted on a special stand built for squareness testing. Once an indicator has been set at zero using the cylindrical square, the dial indicator and the stand can be used to inspect various surface elements of the part surface for a direct reading of the out of squareness of each element, as shown in Figure 9.7. It is important, however, that the indicator move directly vertical during the measurement. The stand must be capable of guiding the indicator vertically.

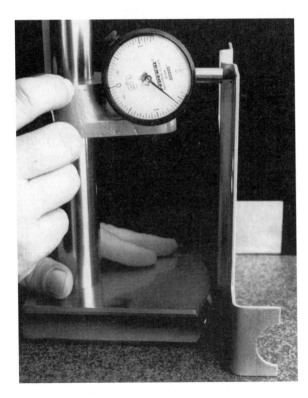

Figure 9.7 Special stand for out-of-squareness inspection using a dial indicator.

PERPENDICULARITY: RIGHT-ANGLE-PLATE METHOD

One of the most accurate (and most popular) methods of inspecting surface-to-surface perpendicularity is the right-angle method (such as a right-angle plate or a precision

130 Chapter 9 Perpendicularity

parallel). The accuracy of this method, however, suffers unless the observer understands the primary problem with using this method. Explanation of the problem begins with a discussion of *alignment* and the need for *secondary alignment datums*. First, the method will be discussed; then the limitation will be covered.

The equipment required for this method is as follows:

1. Surface plate
2. Right-angle plate or precision parallel (for this example, a precision parallel will be used)
3. Two clamps
4. Dial indicator that discriminates to no more than 10% of the part perpendicularity tolerance
5. Surface gage (or height gage) for holding the indicator

The method, when properly performed, re-creates the two parallel planes of tolerance through a FIM reading of the entire controlled surface. The following steps will help the observer make the proper setup for drawings that only call out a primary datum.

STEP 1 Slide the precision parallel and the surface gage onto the surface plate.

STEP 2 Mount the primary datum of the part onto the side of the parallel so that the controlled surface is facing up and somewhat parallel to the surface plate, as shown in Figure 9.8. Clamp the part in a manner that will hold it to the side of the parallel (or a knee), but that will not obstruct the controlled surface.

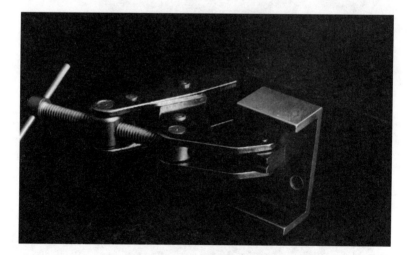

Figure 9.8 The part is mounted and clamped to a precision parallel (on its primary datum).

STEP 3 Prealign the part (especially if there is no secondary datum surface called out for perpendicularity) by *temporarily* bringing a precision square, parallel, or knee against another part surface that is perpendicular to the con-

Perpendicularity: Right-Angle-Plate Method   131

trolled surface (see Figure 9.9). (*Note:* This prealignment is strictly for setup purposes; the second knee, square, or parallel must be removed prior to inspection.)

STEP 4 Find the initial FIM of the surface by placing the indicator at one corner of the controlled surface; then sweep the entire surface (as shown in Figure 9.10).

Figure 9.9 The part is prealigned for the setup using a precision square.

Figure 9.10 The initial FIM is found by sweeping the controlled surface.

132   Chapter 9   Perpendicularity

STEP 5   Rotate the part slightly (left or right) by lightly bumping the part with the hand or a rubber or plastic mallet, and obtain a second FIM on the controlled surface. If the FIM is better than the initial FIM, continue lightly bumping (rotating) the part in that same direction. If the FIM is worse than the initial FIM, rotate the part surface back in the original direction. Continue this rotating and FIM process until *the smallest FIM value is found. The smallest FIM is the perpendicularity error of the part.*

To summarize, the primary problem with this method, as seen in the previous steps, is alignment of the controlled surface when no secondary datum is called out on the drawing. With this method, we can accidentally obtain a wide variety of FIM values, as illustrated in Figure 9.11.

Figure 9.11 illustrates a problem that could have begun in design or in the drafting room manual. Examples in several textbooks on geometric tolerancing show only one datum for perpendicularity of a surface to a surface. This does *not* mean that only one datum is *always* used. A secondary alignment datum for the part should be called out when another part surface (perpendicular to the controlled surface) is also functionally involved (such as a dual interface in assembly). The previous steps must be used when only a primary datum surface is called out. If a secondary datum is identified, the steps change as follows.

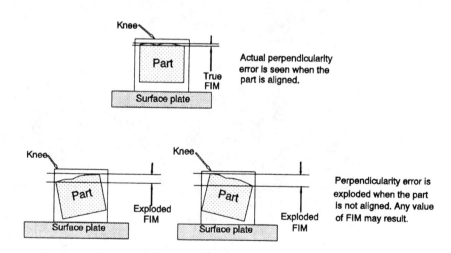

Figure 9.11   Secondary alignment errors when using a precision square (or a right-angle plate) for perpendicularity measurement.

Figure 9.12 changes the drawing to call a secondary datum for the perpendicularity relationship. In this case, the tolerance zone is the same as before *(two perfectly parallel planes that are .005" apart and perfectly perpendicular to datum plane A)*, except that these planes are also *aligned (or perfectly perpendicular to secondary datum plane B)*, as shown in Figure 9.13.

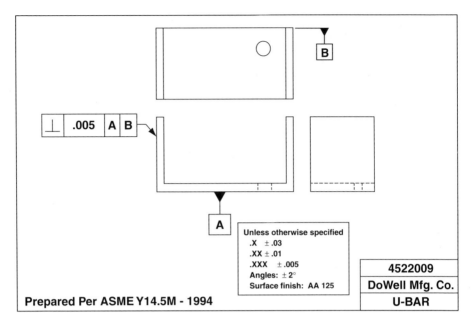

Figure 9.12 Drawing requirement that specifies a secondary alignment datum.

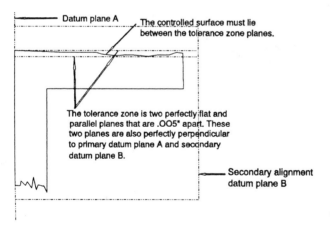

Figure 9.13 Tolerance zone for the part.

In this case, the right-angle plate method is very easy to perform, and *only one FIM* value is needed to inspect the part. The steps for a part with a secondary alignment datum are as follows:

STEP 1 Locate (and clamp) the primary datum surface on one parallel (or knee) to establish a minimum of the three highest points and the secondary datum surface on another parallel (or knee) that is 90° to the first one (to establish the two highest points). Refer to Figure 9.14.

134 Chapter 9 Perpendicularity

Figure 9.14 The part is located on both the primary *and* secondary datums for perpendicularity inspection.

STEP 2 Sweep the entire controlled surface with the dial indicator to obtain a FIM, as shown in Figure 9.15. Rotation of the surface, in this case, is not necessary (and not allowed) because the primary datum supports the part, and the secondary datum aligns the part.

Figure 9.15 One FIM value is all that is needed for perpendicularity inspection when secondary alignment datums are specified.

VERTICAL MEASURING SYSTEMS

A variety of vertical measuring systems are available for measurement of height, centerline measurements, X and Y coordinates, perpendicularity, and other dimensional measurements. These systems often have the appearance of a complicated height gage, and their mode of operation varies, such as operating on dovetail slides, precision ball races, or air bearings.

In cases that require perpendicularity inspection, one thing these systems have in common is a precision vertical movement and a probe or an indicator. Although we would not ordinarily purchase one of these systems just for perpendicularity measurement, they can provide a fast and easy method for measuring perpendicularity if already on hand. An example of this equipment being used to measure perpendicularity is shown in Figure 9.16.

Figure 9.16 Fowler Trimos Vertical 3 Measuring System being used to measure perpendicularity on a part. Courtesy of Fred V. Fowler Co.

PERPENDICULARITY OF AN AXIS (RFS)

Perpendicularity of an axis is often used to control the virtual size of a cylindrical size feature with respect to a datum plane. The size feature, for this example, is a hole that must be perpendicular to a datum plane. The example part is shown in Figure 9.17. Technically, the tolerance zone for perpendicularity of an axis is *two parallel planes of*

136 Chapter 9 Perpendicularity

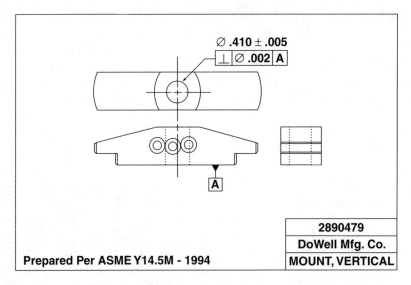

Figure 9.17 Drawing requirement for perpendicularity of a hole to a datum plane.

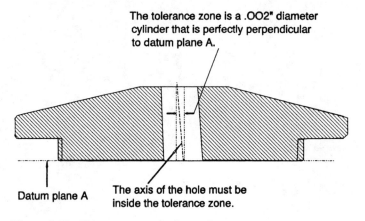

Figure 9.18 Tolerance zone for the part.

infinite rotation (these two planes generate a cylinder); but, for clarity, the standard requires the diameter symbol in the feature control frame. The effective tolerance zone is an *imaginary cylinder with a .002″ diameter that is perfectly perpendicular to primary datum plane A*, as shown in Figure 9.18.

Also, per rule 3 of the standard, if no modifier is shown, RFS applies automatically. In this case, then, the tolerance zone is a true cylinder of the stated tolerance diameter that stands perfectly perpendicular to the datum plane. The axis of the hole must lie completely within the tolerance zone.

Perpendicularity of an Axis (RFS) 137

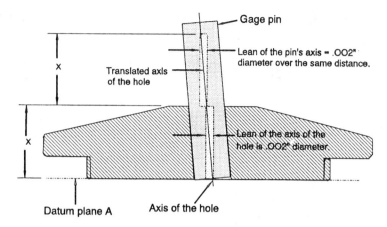

Figure 9.19 The lean of the pin will be the same as the lean of the axis of the hole.

The observer must make an inspection setup which proves that the axis of the hole is in or out of the stated tolerance zone. This same part will also be used in a later MMC example for functional gaging. The following equipment is required to inspect this part in an open setup:

1. Surface plate
2. Right-angle plate (knee) or similar device
3. Dial indicator that discriminates to at least 10% of the total perpendicularity tolerance
4. Surface gage or height gage to hold the indicator
5. Largest true gage pin that will fit inside the controlled hole (and is at least twice the length of the controlled hole)

STEP 1 Carefully slide the inspection equipment onto the surface plate.

STEP 2 Insert the gage pin completely into the hole, but avoid allowing the gage pin to protrude past the datum plane.

STEP 3 Mount the part datum on the right-angle plate, as shown in Figure 9.20, and clamp the part in a manner that will allow access to the gage pin.

STEP 4 At this point, it is important to remember that the *lean* of the axis is being measured, not the location. The pin will follow the lean of the axis of the hole, and so measurements can be made on the pin as long as the measurements are made on a length of the pin equal to the length of the hole. This relationship is shown in Figure 9.19. Measure the lean of the gage pin at top dead center in one location, as shown in Figure 9.20a and b. Then rotate the pin 90° and repeat this step, as shown in Figure 9.21a and b. It is recommended that this step be performed on at least four sections 90° apart.

If the total lean of the pin in the worst direction measured (FIM) is less than or equal to the stated perpendicularity tolerance diameter, the part is acceptable. If the lean

138 Chapter 9 Perpendicularity

Figure 9.20 (a) and (b) Measuring the lean of the pin at one location.

in the worst direction is worse than the stated tolerance diameter, the part should be rejected. (*Note:* Rejections should state the lean in terms of actual diametral zone; for example, the worst lean is .010″ FIM and therefore, in this case, the actual perpendicularity zone is .010″ diameter.)

Keep in mind that, since the tolerance is applied at RFS, the axis of the hole cannot lean out of perpendicularity more than the stated tolerance zone will allow. Since a gage pin was used for the inspection, the observer was able to track the functional axis of the hole by measuring at one edge of the pin. (*Caution:* Do not use this method without a true gage pin because the irregularities of holes produced—for example, circularity, barrel-shaped, and tapered—can cause appreciable error in the measurement.)

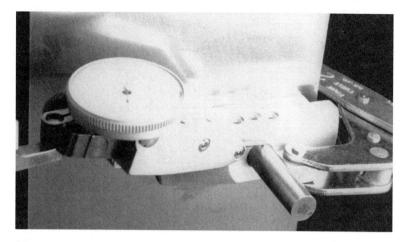

(a)

(b)

Figure 9.21 (a) and (b) Measuring the lean of the pin in another direction.

PERPENDICULARITY OF AN AXIS (MMC)

In certain design applications, the perpendicularity of a hole to a datum pin may be specified at MMC due to the functional relationship of the hole to the mating part and the interface datum. In these cases, a bonus tolerance, if needed, is allowed when the actual size of the controlled hole is larger than the MMC size specified by the drawing. Consider the previous part as redrawn in Figure 9.22. This time, an MMC modifier has been included that requires the hole to be perpendicular to the datum plane within the stated cylindrical tolerance zone *only when the hole is produced at MMC size*. If the

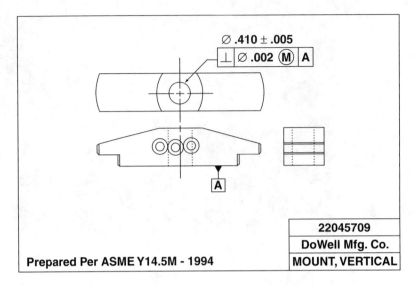

Figure 9.22 Drawing requirement for perpendicularity of a hole to a datum plane at MMC.

actual hole produced is larger than its MMC size, a bonus perpendicularity tolerance can be added to the stated perpendicularity tolerance.

The previous setup can still be used to measure the perpendicularity of the hole (if measurement is required). The only difference is the application of the bonus tolerance at the last step (if the bonus tolerance is needed to make the part acceptable). Another option is to make a *functional gage*. A functional gage is usually a single-purpose gage that is designed to gage the *virtual size* of the part (or, in other words, represent the worst-case mating part). The virtual condition of this hole (allowed by the drawing) is equal to the MMC of the hole minus the position tolerance.

Figure 9.23 provides an example of various hole sizes and perpendicularity errors that may be seen on this part along with the *actual virtual size* of each combination. This table also shows the allowed virtual size of the hole per the drawing. We can see immediately that all the virtual size combinations shown are equal to or larger than the virtual condition allowed. This is why functional gages work. They always gage the worst case, so parts that are better are also accepted.

PERPENDICULARITY OF AN AXIS TO AN AXIS

Another example of perpendicularity of a cylindrical size feature is when the datum is an axis, as shown in Figure 9.24. The feature control frame states that the axes of the controlled holes must be perpendicular to datum axis A within .002″. The tolerance

Actual Hole Size (inches)	Actual Perpendicularity Error (Diameter inches)	Actual Virtual Size (inches)
.405	.002	.403
.406	.003	.403
.407	.004	.403
.408	.005	.403
.409	.006	.403
.410	.007	.403
.411	.008	.403
.412	.009	.403
.413	.010	.403
.414	.011	.403
.415	.012	.403

The virtual size of the part remains constant (.403″) at all combinations of bonus tolerance.

Figure 9.23 Table of examples of actual virtual sizes of the hole versus the allowed virtual condition of the hole.

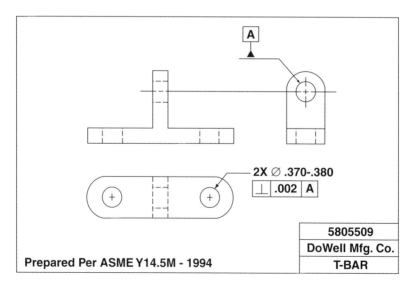

Figure 9.24 Drawing requirement for perpendicularity of the axis of a hole to a datum axis (RFS).

142 Chapter 9 Perpendicularity

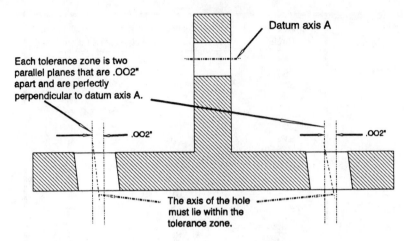

Figure 9.25 Tolerance zone for the part.

zone, for each hole, is *two imaginary and perfectly parallel planes that are .002" apart and perfectly perpendicular to datum axis A*, as shown in Figure 9.25.

To measure the perpendicularity of the hole to the datum on this part, the following equipment is needed:

1. Surface plate
2. Dial indicator that discriminates to no more than 10% of the perpendicularity tolerance
3. Surface gage (or height gage) to hold the dial indicator
4. Gage pin and vee block to contact the datum diameter
5. Largest gage pin that will fit into the controlled holes

The inspection method for perpendicularity of the hole to the datum axis is as follows:

STEP 1 Insert a gage pin into the controlled hole and the datum hole of the part.

STEP 2 Carefully rotate and slide the surface gage and vee block onto the surface plate.

STEP 3 Mount the datum gage pin in the vee block and clamp it down (be careful not to damage the part or the pin during clamping).

STEP 4 Rotate the vee block vertically to make the datum diameter perpendicular to the surface plate, as shown in Figure 9.26a and b.

STEP 5 Indicate the gage pin on both sides of the part, and compare the FIM value to the perpendicularity tolerance, as shown in Figure 9.27a and b. If the FIM value obtained is less than or equal to the perpendicularity tolerance, the part is acceptable. If the FIM exceeds the tolerance, the part is not acceptable.

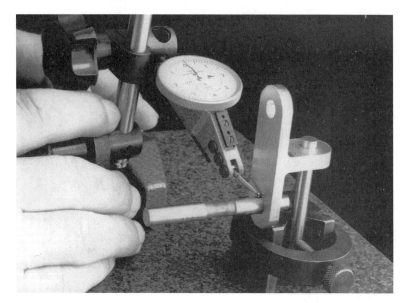

(a)

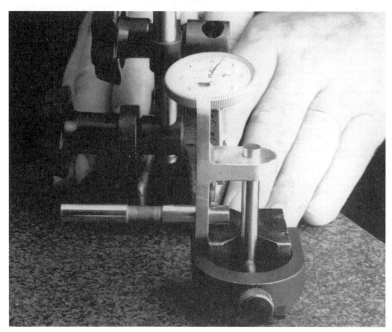

(b)

Figure 9.26 (a) The part is mounted in a vee block where the datum axis is perpendicular to the surface plate. Zero is set at top dead center of the gage pin near the hole. (b) A measurement is taken at top dead center of the gage pin (at the opposite side of the hole).

(a)

(b)

Figure 9.27 (a) and (b) Indicating the gage pin in another direction (at both ends) for perpendicularity error.

Notice that during the measurement there was no concern about the location or direction of the hole—just the perpendicularity of the hole's axis to the datum axis. If the axis of the hole had been perfectly perpendicular, the FIM value would have been zero. The hole, of course, must also be within applicable location tolerances.

REVIEW QUESTIONS

1. Perpendicularity tolerance requires at least one datum reference. *True* or *False?*
2. To which of the following geometric tolerance families does perpendicularity belong?
    (a) Form tolerances
    (b) Location tolerances
    (c) Profile tolerances
    (d) Orientation tolerances
3. What is the first thing an observer should do if a functional gage rejects a perpendicularity tolerance on a hole?
    (a) Put a little more force on the gage pin
    (b) Remount the part on its datums
    (c) Inspect the part by other means
    (d) Reject the gage
4. When only one datum is specified on a surface-to-surface perpendicularity tolerance, the observer needs to consider _____ during inspection.
    (a) Axial location
    (b) Prealignment
    (c) Two highest points on the datum
    (d) Sensitivity of the indicator
5. Perpendicularity of a surface also controls the flatness of the surface within the perpendicularity tolerance zone. *True* or *False?*
6. Perpendicularity tolerances are often applied to two related datum surfaces. The main reason for this is to _____ the datums for use.
7. Which of the following tolerance zones apply to perpendicularity of a surface to a surface?
    (a) Cylindrical zone
    (b) Two parallel planes
    (c) Two parallel lines
    (d) None of the above
8. A perpendicularity tolerance can be inspected using a functional gage if _____ is specified in the feature control frame.
9. What symbol is used in the feature control frame for perpendicularity tolerances that clarifies the tolerance zone for perpendicularity of an axis?
10. What is the shape of the tolerance zone for surface-to-surface perpendicularity when a secondary alignment datum has been identified?
    (a) Two parallel lines
    (b) Cylindrical zone
    (c) Profile zone
    (d) Two parallel planes

10

Angularity

OBJECTIVES

Upon completion of this chapter, the reader should be able to:

1. Explain the difference between simple angle tolerance zones and angularity tolerance zones.
2. Use the sine bar (or sine plate) for angularity inspection.
3. Use angular gage blocks for angularity inspection.
4. Measure the angularity of a conical feature.
5. Explain how to set up a compound sine plate.

INTRODUCTION AND APPLICATIONS

Angularity is an orientation tolerance that provides accurate control over the functional angles of a part (other than simple nonfunctional angles such as clearance chamfers and edge breaks, which do not require the level of control provided by angularity). One of the main applications of angularity is angular interfaces in assembly, or specific angles that have a direct bearing on the fit, form, or function of the part. Angularity tolerance provides a tolerance zone that differs significantly from simple angle tolerances. For example, a simple angle tolerance for a part may be 45° plus or minus 2°. This tolerance (in degrees, minutes, and seconds in some cases) defines a tolerance zone that has an origin and widens over distance (as shown in Figure 10.1).

Angularity tolerance is always specified with respect to a basic angle, a datum reference, and a tolerance zone that is *not* in degrees, minutes, and seconds, but rather in linear units, such as thousandths of an inch or millimeters. The tolerance zone for angularity is *two parallel planes the stated tolerance apart inclined at the basic angle from the datum.*

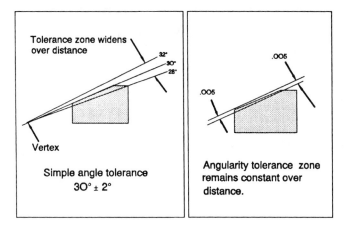

Figure 10.1 The difference between angularity and simple angle tolerance zones.

KEY FACTS

1. Angularity tolerances must be specified with respect to at least one datum reference (at times, more than one is needed).
2. Angularity tolerances require that the desired angle be shown as a basic angle.
3. The angularity tolerance zone in all cases is *two parallel planes* the stated tolerance apart and inclined at the basic angle from the datum reference.
4. Angularity, when applied to a surface, inherently controls the flatness of that surface to the extent of the stated angularity tolerance.
5. Size features controlled with angularity tolerances are automatically regardless of feature size (RFS) applications per rule 3 — unless, maximum material condition (MMC) is stated in the feature control frame.
6. Size features controlled with angularity tolerances are controlled only in the view shown with respect to the datum(s) identified. The feature is not controlled in other directions.

SURFACE-TO-SURFACE ANGULARITY

At times, a surface must be within angularity tolerance to a specified datum feature that is also a surface, such as the part shown in Figure 10.2. The feature control frame states that the surface indicated by the leader arrow must be within a .005″ angularity tolerance to datum surface A with respect to the 30° basic angle. The tolerance zone for this part is *two imaginary planes that are .005″ apart and are inclined at a perfect 30° angle to datum plane A,* as shown in Figure 10.3.

148 Chapter 10 Angularity

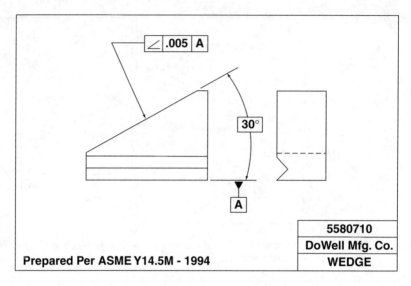

Figure 10.2 Drawing requirement for angularity between two surfaces.

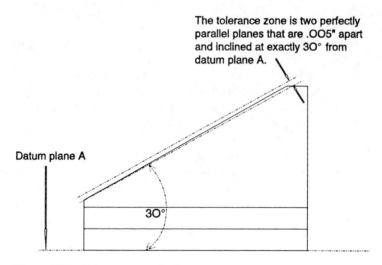

Figure 10.3 Tolerance zone for the part.

Angularity cannot be directly measured with simple instruments such as bevel protractors. The measurements must be made with respect to the perfect (basic) angle in linear units such as thousandths of an inch or millimeters. The result of an angularity inspection is the full indicator movement (FIM) of a dial indicator or probe. The following are popular and effective methods for measuring surface-to-surface angularity.

MEASURING ANGULARITY WITH A SINE BAR OR SINE PLATE

For some time, sine bars and sine plates have been used to measure angularity tolerances directly. The basic operation is simple and uses the sine function of an angle in the setup. A sine bar is a precision bar with two matched cylinders that are a specific distance apart within gage tolerances. For example, a 10″ sine bar is defined by two cylinders that are exactly 10″ apart from center to center within gage tolerances and so on for a 3″, 5″, and 20″ sine bar. These three sizes are typical. The sine bar is always used with a surface plate, gage blocks, and an indicator. The principle of a sine bar is that the two cylinders and their distance apart form the hypotenuse of a right triangle. A specific gage block stack under one end of the sine bar forms the opposite side of a right triangle. When the correct combinations are put together, the surface of the bar is precisely inclined at the selected angle. A simple equation can be used to calculate the amount of gage block stack to put under the stacking end of the sine bar to get the angle desired. This equation is based on the sine function of a right triangle and is shown in Figure 10.4. The sine function of the angle can easily be obtained by using the table of trigonometric functions (see Appendix F) or a scientific calculator. The scientific calculator is a preferred method because of the limitations of tables and the possible errors that can be made reading tables. Almost all calculators use the same function to get the sine of an angle.

First, make sure the calculator mode is in degrees (most calculators will have *Deg* showing in the display). Next, enter the angle (30 in this case) and press the *Sin* key. We can easily test this calculator function because the sine of a 30° angle is always 0.5.

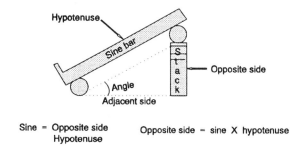

Sine = Opposite side / Hypotenuse   Opposite side = sine × hypotenuse

The sine bar forms the hypotenuse.
The gage block stack forms the opposite side.
The surface plate forms the adjacent side.

Gage stack = sine bar length times the sine of the angle

Figure 10.4 Sine function of a right triangle, and the equation for computing the height of a gage block stack for a sine-bar application.

The example drawing shows that an angle of 30° is required for the angularity tolerance. One example might use a 10″ sine bar to set up the 30° angle for the measurement. If so, the height of the gage block stack that is needed to put under the stack end of the 10″ sine bar is equal to .5 (the sine of 30°) times 10″ equals 5″.

Therefore, a 5″ gage block stack under a 10″ sine bar will tilt the bar exactly 30° from the surface plate. In the following example, a 5″ sine bar will be used for the 30° angle. The equipment needed for the sine bar method is as follows:

1. Surface plate
2. Dial indicator that discriminates to no more than 10% of the angularity tolerance
3. Surface gage (or height gage) to mount the dial indicator
4. Sine bar and clamp
5. Gage block set (or the correct amount of blocks for the application)

STEP 1 Knowing the length of the sine bar and the basic angle of 30° (in this case), you can calculate the gage block stack necessary for the setup. The necessary gage block stack will be the sine function of 30° (which is 0.5) times the length of the sine bar (5″) or *a 2.5″ stack.*

STEP 2 Carefully rotate and slide the surface gage and sine bar onto the surface plate.

STEP 3 Mount the datum of the part onto the surface of the sine bar in a direction that causes the angled surface of the part to be parallel to the surface plate. *(Note: It is important, at this point, to prealign the part on the sine bar since no alignment datum is called out on the drawing.* Recall the problems that were discussed on secondary alignment datums in Chapter 9. The same problems occur in this method without alignment datums.) To prealign the part on the sine bar, use any precision-square accessory (such as a knee, 1–2–3 block, gage block, or precision parallel) and align the part with the side of the bar, as shown in Figure 10.5. (*Note:* Some sine bars have an end plate mounted on them that can be used to prealign the part. The part, however, cannot be mounted against the end plate during the inspection if the drawing does not specify a secondary datum. Also note that light clamping may be necessary if the part will not remain stable under its own weight.)

STEP 4 Carefully rotate and slide the gage block stack onto the surface plate and lift the stack end of the sine bar. Place the gage block stack under the cylinder at the stack end of the sine bar, as shown in Figure 10.5. (*Note:* The stack end of a sine bar is usually the lighter end or the end on some sine bars that does not have a back stop on it.)

STEP 5 Obtain an initial FIM value while the part is prealigned. This value may change due to alignment, but it will provide a good reference value. See Figure 10.6.

STEP 6 In this step, the measurement of angularity is made, but the lack of a secondary alignment datum makes measurement somewhat tedious. The goal of this step is to establish an FIM of the controlled surface *when it is aligned.* First, sweep the controlled surface as it has been set up and establish an FIM

Measuring Angularity with a Sine Bar or Sine Plate   151

Figure 10.5 The part is mounted on the sine bar and prealigned with a precision parallel.

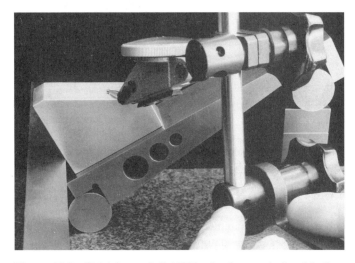

Figure 10.6 Obtaining an initial FIM value for angularity while the part is prealigned.

value, as shown in Figure 10.7. Next, rotate the part left and right slightly, obtaining other FIM values each time. The *smallest* FIM value is the angularity of the actual surface. If the smallest FIM value is equal to or less than the angularity tolerance, the part is acceptable. If the smallest FIM value is larger than the angularity tolerance, the part is not acceptable.

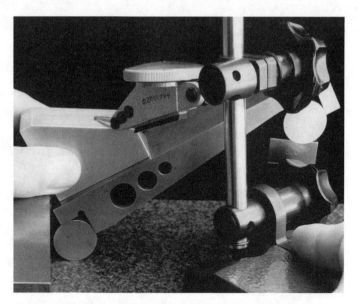

Figure 10.7 Searching for the smallest FIM value (because a secondary alignment datum has not been specified).

If the angularity of the part had been perfect, the smallest FIM value would have been zero. The observer should take caution when measuring angularity of a surface without a secondary alignment datum specified on the drawing. Appreciable errors can result if we simply set up the part on the sine bar and take a random FIM value.

SINE-BAR METHOD: ANGULARITY WITH SECONDARY DATUMS

Figure 10.8 is an example of an angularity tolerance that has been specified to a primary *and* a secondary datum. The tolerance zone is *two parallel planes that are .005" apart and inclined at a perfect 30° angle to datum plane A* and *aligned by secondary datum plane B,* as shown in Figure 10.9. This part is much easier than the previous part to inspect using a sine bar. The equipment for this method is the same as before except that a parallel, right-angle plate, for example, is used to contact the secondary datum. Again, some sine bars (and sine plates) have built-in end plates for this purpose. The steps for inspection are the same except that the part is mounted and clamped against both datums (primary first on the bar, and secondary next using the parallel), and *only one FIM value is needed. The part is not rotated for the smallest FIM value.*

Sine-Bar Method: Angularity with Secondary Datums

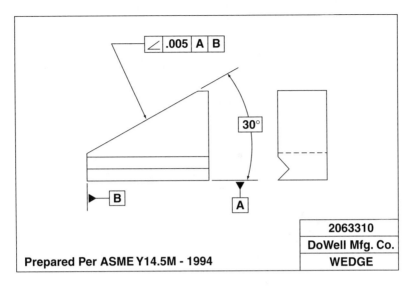

Figure 10.8 Drawing requirement for angularity that specified both primary and secondary datums.

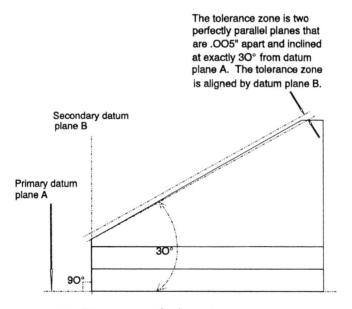

Figure 10.9 Tolerance zone for the part.

SINE PLATES

A sine plate (setup shown in Figure 10.10) works the same way as the sine bar and holds larger (wider) part datums. Sine plates are often the best choice because sine bars are limited to the selected sizes of parts that they can hold. Remember, the primary datum (for a surface) is the three (or more) highest points on that surface. The sine plate or bar must contact the entire datum surface of the part to establish the datum plane.

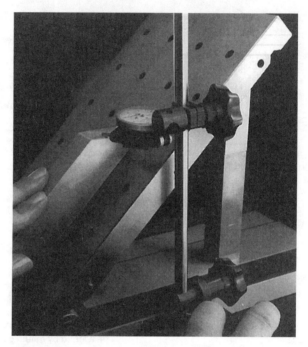

Figure 10.10 Sine plate being used for angularity inspection.

ANGULAR-GAGE-BLOCK METHOD

Angular-gage-block sets can also be used to set the basic angle for angularity inspection, as shown in Figure 10.11. Angular gage blocks are as accurate in their angle as gage blocks are accurate in their length. The accuracies of angular gage blocks can range from working blocks that are accurate to plus or minus 1 second of arc to lab-grade blocks that are accurate to plus or minus 1/4 second. Angular gage blocks come in three different sets. Each of the following sets has an angle range from 0° to 90°:

1. Six-block set; the smallest increment of angle that can be produced is 1°.
2. Eleven-block set; the smallest increment of angle that can be produced is 1 minute.
3. Sixteen-block set; the smallest increment of angle that can be produced is 1 second.

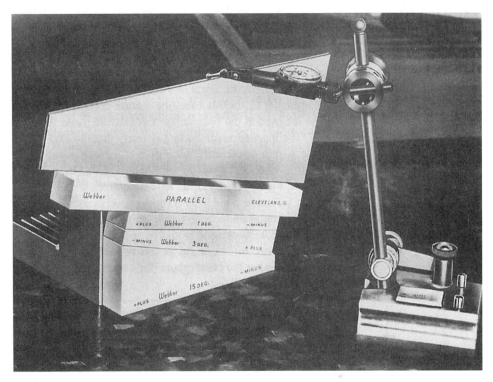

Figure 10.11 A part is being inspected for angularity using angled gage blocks, a right-angle plate, a precision parallel, and a dial indicator. Courtesy of L.S. Starrett Co.

Thousands of different angles can be established using set 3 because the blocks are designed in a way that they can be made to add or subtract to the stack (or combination). An example of angular-gage-block stacks is shown in Figure 10.12.

To use angular gage blocks for angularity inspection, they often must be stabilized against a right-angle plate or precision parallel so that the part datum can rest on the angled surface. A limitation of angular gage blocks is the fact that many part datums are too wide to rest on the limited area of the block.

It is important to remember that the primary datum for the part angularity requirement must be fully contacted to establish the three (or more) highest points of the datum surface. It is for this reason that sine bars and sine plates can be a more appropriate tool for the job.

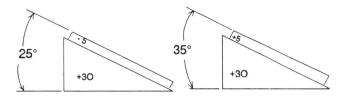

Figure 10.12 Examples of angular-gage-block stacks.

ANGULARITY OF A SIZE FEATURE

In other design applications for angularity, size features (for example, holes, pins, tabs, or bosses) must be controlled at a specific angle to a datum feature. An example of this application is shown in Figure 10.13. In this case, the feature control frame states that the axis of the hole must be inclined at the basic angle from datum plane A within an angularity tolerance of .002". The inspection method must verify that the axis of the hole is inclined to the basic angle within a tolerance zone of *two parallel planes that are .002" apart and are inclined at a basic 20° from datum plane A*. Even though the controlled hole is a cylindrical feature, the angularity of its axis is controlled by a tolerance zone of two planes. The angularity tolerance zone is RFS (understood by rule 3), and so no bonus tolerance is allowed. It is also important to note at this point that if the hole were a pin, a boss, or a slot, the angularity tolerance zone would be the same two parallel planes. The inspection method, in these cases, must use equipment or evaluation to find the inclination of the axis (for a hole or pin) or the centerplane (for bosses and slots).

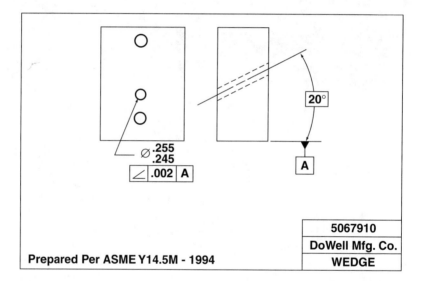

Figure 10.13 Drawing requirement for angularity of an axis of a hole to a datum plane.

ANGULARITY OF A HOLE: SINE-BAR METHOD

The sine bar is also a very good method for inspection of the angularity tolerance in Figure 10.13. The tolerance zone for angularity of the hole is *two parallel planes that are .002" apart and inclined at a perfect 20° from datum plane A*, as shown in Figure 10.14.

When assisted with the largest true gage pin that can be inserted into the controlled hole, the sine bar can, as before, establish the basic angle in a manner such that the pin

Angularity of a Hole: Sine-Bar Method

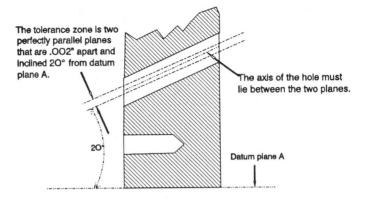

Figure 10.14 Tolerance zone for the part.

(inserted into the hole) would be parallel to the surface plate if the hole were inclined at the basic angle. If not, the FIM reading on the dial indicator would show the amount of angularity error to be compared to the stated tolerance. Notice, in this example, that the same sine bar is used that was used in the earlier surface-to-surface example, but the angle of the hole has been changed. The inspection equipment needed for this measurement is as follows:

1. Surface plate
2. Dial indicator that discriminates to no more than 10% of the angularity tolerance
3. Surface gage (or height gage) to hold the dial indicator
4. Appropriate gage block stack for the angle and length of the sine bar
5. Sine bar
6. Largest gage pin that can be inserted into the controlled hole

STEP 1 Knowing the basic angle of the hole (20°) and the length of the sine bar (5″), you can calculate the appropriate gage block stack that will raise the sine bar at the 20° angle. The gage block stack needed is equal to the sine of 20° (.34202) times the length of the bar (5″), or 1.7101″.

STEP 2 Insert the largest gage pin into the controlled hole. The angularity inspection must be made in a distance of the pin that is equal to the depth of the hole. If the hole were drilled through the part, the pin should be inserted in a manner such that it sticks out at each end of the hole, and the measurements are made at the entrance and exit locations on the pin.

STEP 3 Carefully rotate and slide the surface gage, gage block stack, and sine bar onto the surface plate. Place the gage block stack under the stack end of the sine bar to set up the basic angle for measurement, as shown in Figure 10.15.

STEP 4 Mount the datum of the part onto the surface of the sine bar in a direction that causes the controlled hole to be parallel to the surface plate. (*Note: Once again it is important to prealign the part on the sine bar since no alignment datum is called out on the drawing.*)

158 Chapter 10 Angularity

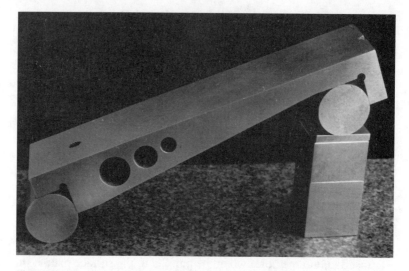

Figure 10.15 The sine bar has been set up at the basic angle for the measurement.

To prealign the part on the sine bar, use any flat surface (such as a gage block or precision parallel) and align the part with the side of the bar, as shown in Figure 10.16, or use the end plate that is mounted on some sine-bar models to align the part. The part, however, cannot be mounted against the end plate during the inspection if the drawing does not specify a secondary datum.

Figure 10.16 The part datum is mounted on the sine bar, and the part is prealigned using a precision parallel.

Angularity of a Hole: Sine-Bar Method

STEP 5 In this step, the measurement of angularity is made, but the lack of a secondary alignment datum once again makes measurement somewhat tedious. The goal of this step is to establish an FIM of the controlled surface *when it is aligned*. First, sweep the surface of the pin as it has been set up and establish an FIM value, as shown in Figure 10.17a and b. Next, rotate the part left and right slightly, obtaining other FIM values each time, as

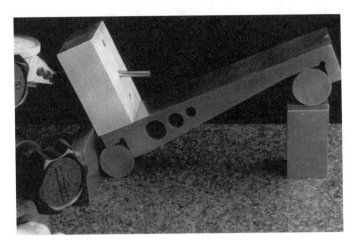

(a)

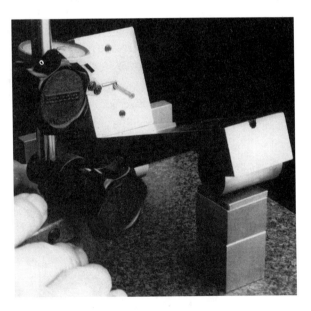

(b)

Figure 10.17 (a) and (b) Obtaining an initial FIM value for angularity.

Figure 10.18 Obtaining the best FIM value for angularity.

shown in Figure 10.18. The *smallest* FIM value is the angularity of the axis of the hole. If the smallest FIM value is equal to or less than the angularity tolerance, the part is acceptable. If the smallest FIM value is larger than the angularity tolerance, the part is not acceptable.

If the angularity of the axis had been perfect, the smallest FIM value would have been zero. The observer should take caution when measuring the angularity of a feature without a secondary alignment datum specified on the drawing. Appreciable errors can result if we simply set up the part on the sine bar and take a random FIM value.

ANGULARITY OF A CONE

In other cases, as in the example in Figure 10.19, the angularity of a cone is an important functional requirement of a part. The feature control frame requires that the surface elements of the cone be within a .001″ angularity tolerance with respect to datum axis A. The basic angle that locates this tolerance zone is 10°. The tolerance zone is *two imaginary lines that are .001″ apart and inclined at a perfect 10° from datum axis A,* as shown in Figure 10.20.

Each line element of the cone must lie within the tolerance zone. In these cases, the angularity tolerance will be specified with respect to a basic angle that is either the included angle of the cone (not shown) or one-half the included angle, as shown on the drawing. Sine bars work on a per-side basis, and so if the included angle were shown, the observer would use one-half the included angle to compute the gage block stack.

Angularity of a Cone

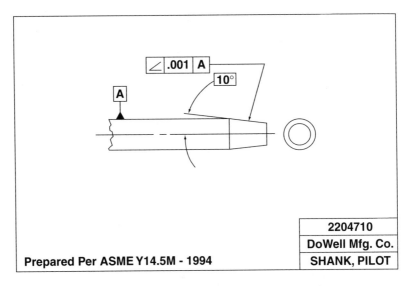

Figure 10.19 Drawing requirement for angularity of a cone.

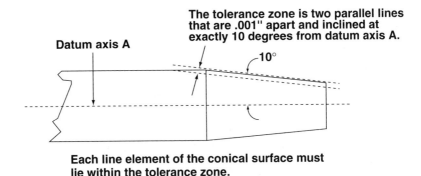

Figure 10.20 Tolerance zone for the part.

The inspection equipment needed for this measurement is as follows:

1. Surface plate
2. Dial indicator that discriminates to no more than 10% of the angularity tolerance
3. Surface gage (or height gage) to hold the dial indicator
4. Sine bar
5. Appropriate gage block stack
6. Vee block or magnetic vee block (as shown in the following figures)

STEP 1 Knowing the basic angle of the cone (per side) and the length of the sine bar (5″), you can calculate the appropriate gage block stack that will raise the sine bar at the basic angle.

STEP 2 Carefully rotate and slide the surface gage, gage block stack, magnetic block (or vee block), and sine bar onto the surface plate. Place the gage block stack under the stack end of the sine bar to set up the basic angle for measurement, as shown in Figure 10.21.

STEP 3 Mount and clamp the part into the vee block. The vee block is used to hold the part onto the sine bar.

STEP 4 Mount the vee block and part onto the surface of the sine bar in a direction that causes the inclined edge of the cone to be parallel to the surface plate. (*Note: Once again it is important to prealign the part on the sine bar since no alignment datum is called out on the drawing.*)

To prealign this part on the sine bar, use any square accessory (such as a gage block, 1-2-3 block, knee, or precision parallel) and align the part with the side of the bar, as shown in Figure 10.22, or use the end plate that is

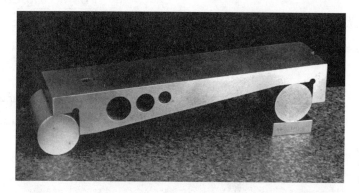

Figure 10.21 The sine bar has been set up at the basic angle of the cone.

Figure 10.22 The part and vee block are mounted on the sine bar and prealigned using a precision parallel.

Angularity of a Cone

(a)

(b)

Figure 10.23 (a) and (b) Obtaining an initial FIM value for angularity of the cone.

mounted on some sine-bar models to align the part. The part, however, cannot be mounted against the end alignment plate during the final step of inspection if the drawing does not specify a secondary datum.

STEP 5 A lack of a secondary alignment datum once again makes measurement somewhat tedious. The goal of this step is to establish an FIM of the controlled line element of the cone *when it is aligned*. First, sweep the controlled

164   Chapter 10   Angularity

line element as it has been set up and establish an FIM value, as shown in Figure 10.23a and b. Next, rotate the part left and right slightly, obtaining other FIM values each time, using the same method as shown in Figure 10.23a and b. The *smallest* FIM value is the angularity of a line element of the cone.

STEP 6   Rotate the cone to another line element of the cone and repeat step 5 once again. If the smallest FIM value for each line element of the cone is equal to or less than the angularity tolerance, the part is acceptable. If the smallest FIM value is larger than the angularity tolerance, the part is not acceptable. If the angularity of a line element of the cone had been perfect, the smallest FIM value would have been zero.

SETTING COMPOUND SINE PLATES

Some parts have compound angles that are 90° to each other. To inspect the part, a compound sine plate is often necessary. Compound sine plates are not as straightforward as simple sine plates and sine bars. Setting them requires a specific procedure. Refer to Figure 10.24 for the example and assume that the first angle must be 30° and the second, 20°.

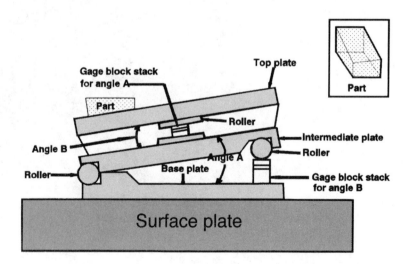

Figure 10.24   Setting up a compound sine plate for a compound angle.

STEP 1   Look up the cosine of the second angle and the tangent of the first angle, and multiply them.

$$\cos_{20} \times \tan_{30} = \text{Answer}$$

STEP 2   Take the product of step 1 and find it under *tangents* in a table of trigonometric functions, and then find the *sine* equivalent of that value.

STEP 3 Multiply the *sine* times the length of the compound sine plate (5" or 10"). This is the gage block stack that must be placed under the intermediate plate to establish angle A. Wring the proper gage block stack for angle A, and set it between the base plate and the intermediate plate.

STEP 4 Look up the sine of the second angle B, and multiply the sine times the length of the compound sine plate (5" or 10"). Wring a gage block stack and set it between the intermediate plate and the top plate to establish angle B.

Sine bars, sine plates, and compound sine plates are very useful for inspection of angularity tolerances on a wide variety of component parts.

UPSIDE-DOWN USE OF THE SINE BAR

Another way of using the sine bar for measuring tightly toleranced angles is the *upside-down* sine bar method. In this application, the working surface of the sine bar is set on top of the controlled surface of the part, and a height measurement is made between the matched cylinders of the sine bar. The sine formula is then used to calculate the angle of the workpiece. An example is shown in Figure 10.25.

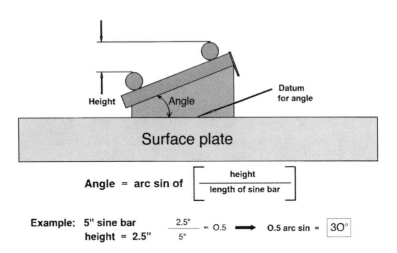

Figure 10.25 Upside-down use of a sine bar for measuring angles.

The inspection equipment needed for this measurement is as follows:

1. Surface plate
2. Sine bar (any length)
3. Dial indicator that discriminates to 10% or less of the tolerance
4. Height gage to hold the dial indicator and make height measurements

STEP 1 Carefully put the part on the surface plate (mounted on the primary datum).

STEP 2 Place the sine-bar mounting surface on the angled surface of the part. Make sure it is aligned. (*Note:* Alignment should not be a problem if the sine bar has an end plate.)

STEP 3 With the height gage and dial indicator, measure the distance from the top of one sine-bar cylinder to the top of the other.

STEP 4 Using the length of the sine bar and the measured distance between the matched sine-bar cylinders, calculate the angle of the workpiece using the equation in Figure 10.25. Note that the height between the cylinders divided by the length of the sine bar results in the sine function of the angle. The observer must arc the sine function to obtain the angle of the workpiece. For example, the sine function of 20° is .3420 (rounded to four decimal places). Therefore, the arc sine of .3420 is equal to 20°. Any scientific calculator can be used by simply entering (for example) ".3420 Inv Sin" or ".3420 2nd Sin," depending on the calculator button that is used for the second function of the sine key.

If the resultant angle is within the angular tolerance, the part is acceptable. It is important to note that, using this method, an FIM value cannot be directly obtained for angularity tolerances. Those boundaries would have to be calculated using right triangle solutions. However, the precise angle of the workpiece will be known.

REVIEW QUESTIONS

1. Angularity can be inspected using a bevel protractor. *True* or *False?*
2. What is the height of the gage block stack required to set up a 30° angle on a 10″ sine bar (or sine plate)?
3. Which of the following best describes the difference between angularity and simple angle call-outs?
   (a) Angularity requires a specified datum.
   (b) Angularity tolerance zones do not widen over distance.
   (c) Angularity is an FIM value, not degrees, minutes, and seconds.
   (d) All of the above
4. What is the shape of the tolerance zone for angularity of a cone?
   (a) Two parallel planes inclined at the basic angle
   (b) Two parallel lines inclined at the basic angle
   (c) A cylindrical zone inclined at the basic angle
   (d) None of the above
5. When inspecting angularity to a primary datum on a sine bar (or sine plate), the observer must always be aware of a possible cause of error in the FIM value due to _____.
6. What is the sine of 60° rounded off to three decimal places?
7. Angularity tolerance is a member of the form tolerance family. *True* or *False?*

8. What is the shape of the tolerance zone in cases where angularity tolerances are applied to the axis of a hole?
   (a) A cylinder
   (b) Two parallel lines
   (c) Two parallel planes
   (d) Angularity cannot be applied to an axis
9. Which of the following choices of surface-plate accessories can be readily used to inspect angularity?
   (a) 1-2-3 blocks
   (b) Rectangular gage blocks
   (c) Precision parallels
   (d) Angular gage blocks
10. Name one of the following that does not apply to angularity: (a) datums, (b) FIM, (c) vertex, (d) basic angles.

11

Circular Runout

OBJECTIVES

Upon completion of this chapter, the reader should be able to:

1. Inspect circular runout on a diameter or an end surface.
2. Inspect circular runout on a cone.
3. Inspect circular runout with respect to offset compound datum diameters.

INTRODUCTION AND APPLICATIONS

Certain geometrical errors have a direct impact on the function of rotating parts. Rotating parts and assemblies almost all have one important function in common: They must rotate with minimal vibration. Vibration in rotating parts creates unwanted noise and, at times, early failures. Vibration in rotating assemblies can be caused by unbalanced rotating parts. Unbalance is created when too much mass (material) has deviated unevenly from a rotating axis. Geometrical errors that contribute to unbalance of rotating parts (and subsequently rotating assemblies) are roundness (circularity) error, eccentricity, and at times, perpendicularity error. Runout tolerance is designed specifically to control all these errors collectively.

Circular runout is a surface-to-datum axis control that is applied to rotating parts with respect to at least one datum reference. Due to the circular runout tolerance zone and its application on a diameter, there is inherent control over circularity, concentricity, and wobble of the diameter. When a circular runout tolerance is applied to an end surface, it inherently controls the wobble and perpendicularity of that surface. The usual functional applications of runout tolerances are related to rotating parts for functional purposes such as balance or concentricity control or other controls having to due with rotating mass.

KEY FACTS

1. Circular runout must be specified with respect to at least one datum reference at all times. The datum feature(s) establish(es) a datum axis of rotation.
2. The measurement of circular runout requires independent FIM values obtained at various circular elements while the part is rotating about the datum axis 360°. Each circular element is inspected separately.
3. Circular runout inherently controls circularity and concentricity because the cumulative variations of these two geometric errors are seen in the surface-to-axis relationship.
4. Runout tolerances are always regardless of feature size (RFS) and are evaluated on a full indicator movement (FIM) basis.
5. Circular runout can be applied to conical features to control roundness, concentricity, and runout of each circular section of the cone. Measurements of circular runout on conical features must be made with the indicator or probe positioned at right angles to the conical surface.

CIRCULAR RUNOUT: VEE-BLOCK METHOD

Vee blocks have been used for many years to locate outside-diameter datum features for measurements of runout, concentricity, and other applications. The use of the vee block, however, is not that straightforward. It is important that the observer understand the limitations and possible errors that can result when using vee blocks. First, from a technical standpoint, vee blocks only contact *two line elements* of the diameter. This type of datum feature contact is normally related to a drawing callout of datum target lines accompanied by a basic angle (as shown in Figure 11.1).

Although the callout in Figure 11.1 is rarely seen, vee blocks are still used in industry to contact datum diameters for measurement. The most important aspect of using vee blocks is to make sure that the datum diameter being located is *qualified* so that the two lines of contact will generate an axis that is closely coincident with the true axis of the diameter. Consider the drawing with a runout requirement in Figure 11.2, in

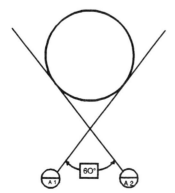

Figure 11.1 Drawing requirement for two line elements of contact on a diameter at a specific angle apart.

170 Chapter 11 Circular Runout

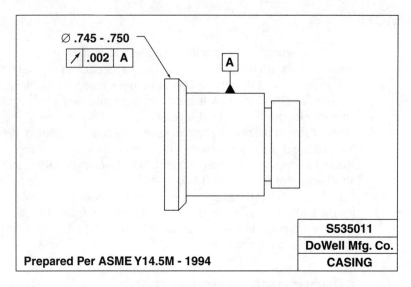

Figure 11.2 Drawing requirement for circular runout.

which each circular element of the largest diameter must not run out from datum axis *A* more than .002″ (FIM is understood). The tolerance zone for circular runout of the surface of a diameter is, in effect, *two points at each circular element that are .002″ apart,* as shown in Figure 11.3.

If datum *A* is qualified (refer to Chapter 2 discussion), a vee block can be used to locate datum feature *A* with negligible error. The following equipment is needed for the vee-block method of circular runout measurement:

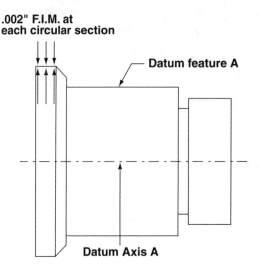

Figure 11.3 Tolerance zone for the part.

1. Surface plate
2. Vee block large enough to accept datum diameter A
3. Dial indicator that discriminates to no more than 10% of the runout tolerance
4. Surface gage (or height gage) to mount the dial indicator

STEP 1   Carefully slide the vee block and surface gage onto the surface plate.

STEP 2   Locate the datum A diameter in the vee block, as shown in Figure 11.4.

STEP 3   Adjust the surface gage so that the indicator tip makes contact near top dead center of the controlled diameter and there is enough travel on the indicator for the runout measurement.

STEP 4   Rotate the part 360° in the vee block (on datum A) and look for the FIM value. The FIM value should not exceed the tolerance value shown on the drawing.

STEP 5   Move the dial indicator over slightly (in the direction of the axis of the controlled diameter) and repeat step 4. This FIM value should also be equal to or less than the stated drawing tolerance for runout. (*Note:* Normally, the observer would repeat these steps on several circular elements of the controlled diameter, but this particular part has such a short axis that two or three elements will usually suffice.)

If all elements inspected do not exceed the .002″ runout tolerance (FIM), the part is acceptable. If *any* single element exceeds .002″ FIM, the part is not acceptable. In

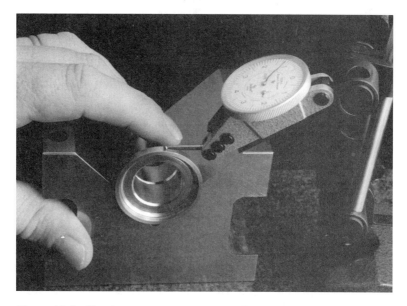

Figure 11.4   Circular runout measurement using a vee block and a dial indicator.

172   Chapter 11   Circular Runout

summary, circular runout on a diameter can be affected by circularity error, eccentricity, and other geometric errors of the controlled diameter or the datum diameter. All these errors are involved and reflected in the FIM value obtained.

It is important to make sure that datum diameters are qualified for use in vee blocks. A good rule of thumb is that the error of the datum diameter should not exceed 10% of the runout tolerance.

CIRCULAR RUNOUT: INSIDE-DIAMETER DATUMS

In many cases, a circular runout requirement may be identified with respect to inside-diameter datum features that establish the datum axis. In these cases, it is best when a true gage pin can be inserted into the datum hole to establish the axis of rotation for the measurement. Gage pin wires can be used to contact datum inside diameters on small holes. Once the pins have been inserted, two methods can be used to find the axis of rotation. These are matched vee blocks or adjustable centers. For adjustable centers to be used, the pin wire must have centers. In either case, it is best to find the largest true gage pin that will fit in the datum diameter to establish a datum axis. Refer again to Chapter 2 discussion on simulated datums for further information.

Consider the drawing in Figure 11.5. The runout requirement is an outside diameter that must not exceed a circular runout tolerance of .005″ with respect to the inside diameter (datum *A*). The equipment required for this inspection method is as follows:

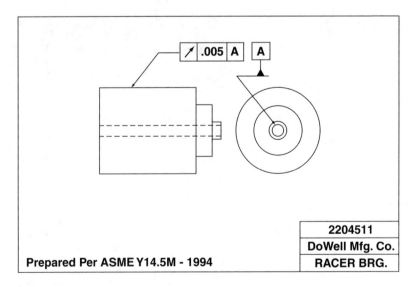

Figure 11.5   Drawing requirement for circular runout when the datum feature is an inside diameter.

1. Surface plate
2. Surface gage (or height gage)
3. Dial indicator that discriminates to no more than 10% of the runout tolerance
4. Set of matched vee blocks (or adjustable centers)
5. Largest true gage pin that will fit into the hole (datum *A*). (*Note:* If the gage pin has centers, adjustable centers or matched vee blocks can be used. For this example, we will use matched vee blocks.)

STEP 1   Carefully slide the surface gage and matched vee blocks onto the surface plate.

STEP 2   Insert the gage pin into datum *A* on the part.

STEP 3   Position the matched vee blocks at a distance apart so that the part will go between them and the gage pin will rest in both blocks (as shown in Figure 11.6).

STEP 4   Position the part between the matched vee blocks, as shown in Figure 11.6.

STEP 5   Position the dial indicator at one end of the controlled diameter and at top dead center. Make sure that there is enough travel on the indicator for the measurement.

STEP 6   Rotate the part 360° and watch for the FIM on the indicator. Move the indicator over to another circular element and repeat this step. For best results,

Figure 11.6 Setup for runout inspection using a gage pin and a matched vee-block set.

several circular elements should be evaluated separately. At least three FIM readings should be taken (one end, the middle, and the opposite end of the controlled diameter). If each independent FIM value is equal to or less than the circular runout tolerance, the part is acceptable. If any circular element exceeds the runout tolerance, the part is not acceptable.

In summary, the circular runout requirement applies to dependent circular elements of the controlled diameter, and if only one area exceeds the runout tolerance, this is a basis for rejection of the part. With this method, both hands are free to concentrate on rotating the part in the matched vee blocks.

END-SURFACE CIRCULAR RUNOUT: TWO DATUMS

Circular runout can be applied to an end surface with respect to a datum axis. In this case, two geometric errors (wobble and perpendicularity of the end surface) are automatically controlled. The collective effect of perpendicularity and wobble is reflected in the FIM value obtained from the end-surface runout measurement. Consider the drawing in Figure 11.7, which requires that the circular runout on the end surface of the part shall not exceed .001″ FIM with respect to the axis generated by datums *A* (primary) and *C* (secondary). The equipment necessary for this inspection method is as follows:

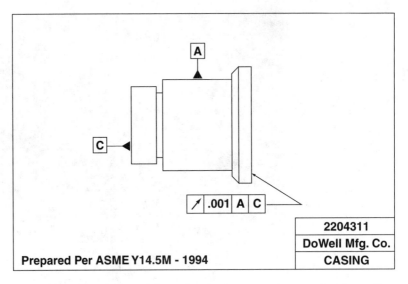

Figure 11.7 Drawing requirement for circular runout on an end surface with respect to two datums.

1. Surface plate
2. Surface gage (or height gage)
3. Dial indicator that discriminates to no more than 10% of the runout tolerance

4. Vee block (to contact datum *A*)
5. Another block (gage block, precision parallel, or other) to mount in the vee block for contacting datum *C*
6. Clamp, if necessary, to clamp the gage block onto the vee block

STEP 1   Carefully slide the vee block and surface gage onto the surface plate.

STEP 2   Mount datum diameter *A* in the vee block such that the entire diameter is located in the vee block.

STEP 3   Mount a gage block (or precision parallel) in the vee block, as shown in Figure 11.8. The gage block rests inside the vee angle of the vee block in a manner such that the side of the gage block is perpendicular to each wall of the vee angle, as shown in Figure 11.8. The gage block is mounted after the *A* datum is located into the vee block, because the end face *C* of the part is the secondary datum.

STEP 4   Place the indicator tip at one location on the controlled end face of the part and, holding the part down in the vee block and back against the gage block, rotate the part 360°. While rotating the part, watch for the FIM value, as shown in Figure 11.8. The FIM should not exceed the stated runout tolerance on the drawing.

STEP 5   Normally, on wider end faces, the indicator would be moved over slightly to another circular element, and step 4 would be repeated. In this particular case, the land width of the end face being controlled is so narrow that the

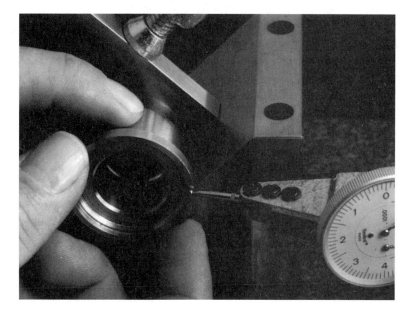

Figure 11.8   Inspecting the part.

observer would only need to move the indicator slightly and repeat step 4 in one more area.

The FIM value, as before, cannot exceed the stated runout tolerance at any circular element inspected. In cases similar to the part that was just inspected, end-surface runout was not a major problem because of the secondary alignment datum that was used. In some cases, end-surface runout callouts do not have a secondary alignment datum, and the part is nearly impossible to inspect in a vee block because the part slides back and forth along its axis and throws error into the FIM value found by the indicator. In such cases, we must contact the primary diameter in a manner that will establish the axis but not allow the datum diameter to slide axially. A precision spindle or similar equipment will meet this requirement because the datum diameter is clamped during rotation. An example of this setup is shown in Figure 11.9.

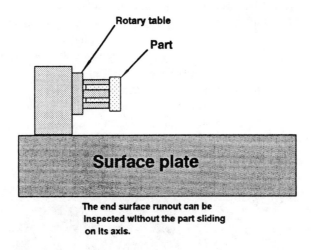

Figure 11.9 A rotary table is used to inspect end-surface runout with respect to a primary datum axis.

CIRCULAR RUNOUT APPLIED TO A CONE

Circular runout is the only runout tolerance that can be applied to a cone because circular runout applies only to independent circular sections of a diameter. Total runout, as discussed in Chapter 12, cannot be used on conical sections of diameters because the composite FIM required for total runout does not allow taper beyond the tolerance band. An example of circular runout on a cone is shown in Figure 11.10. The feature

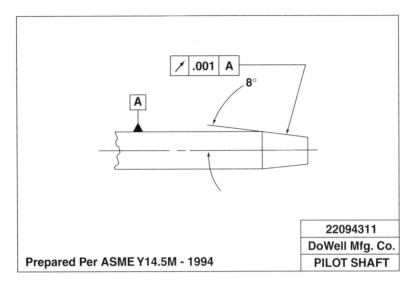

Figure 11.10 Drawing requirement for circular runout on a cone.

control frame states that each circular element of the cone must not exceed a circular runout requirement of .001" (FIM is understood) with respect to primary datum axis A.

The following equipment is necessary for this inspection method:

1. Surface plate
2. Dial indicator that discriminates to no more than 10% of the circular runout tolerance
3. Surface gage to hold the dial indicator
4. Vee block

STEP 1 Carefully rotate and slide the equipment onto the surface plate.

STEP 2 Mount the primary datum diameter in the vee block and on the sine bar, as shown in Figure 11.11. The primary purpose for the sine bar is to ensure that the indicator is at a right angle to the cone surface.

STEP 3 Position the dial indicator at one end of the cone, as shown in Figure 11.12.

STEP 4 Rotate the part in the vee block, and watch for the FIM value obtained at one circular element, as shown in Figure 11.13. Move the dial indicator to another circular element, position the indicator once again, and obtain another FIM value. If all FIM values are less than or equal to the runout tolerance, the part is acceptable. If *any* FIM value exceeds the runout tolerance, the part is not acceptable.

Chapter 11 Circular Runout

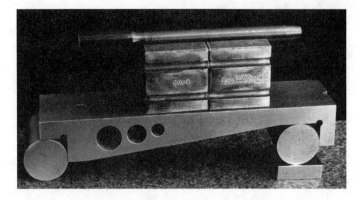

Figure 11.11 The part is mounted and ready for inspection.

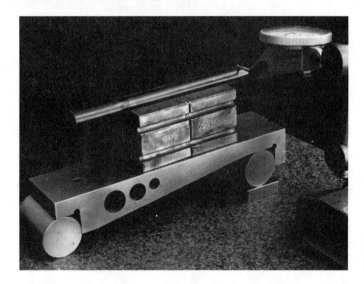

Figure 11.12 The dial indicator is positioned at one end of the cone for runout inspection.

Figure 11.13 FIM values are obtained at various circular elements of the cone.

CIRCULAR RUNOUT: COMPOUND OFFSET DATUMS

Another example of circular runout to compound datums is shown on the drawing in Figure 11.14. In this case, a diameter of the shaft must not run out more than .003″ to compound datum diameters *B* and *C*. Since both datum diameters share the same axis and are a different size, the vee-block method for inspecting this part would have to offset the vee blocks to establish the datum axis. We have the choice to using a matched set of vee blocks and gage blocks to provide the proper amount of offset or the adjustable vee blocks shown in Figure 11.15.

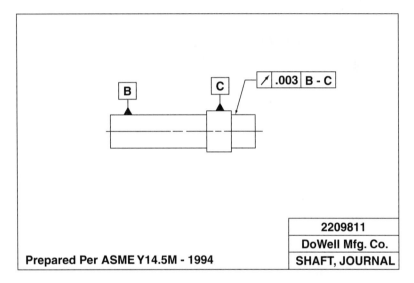

Figure 11.14 Drawing requirement for circular runout with respect to multiple offset datum diameters.

The equipment required for open-setup inspection of this runout requirement is as follows:

1. Surface plate
2. Pair of matched vee blocks and a gage block set (or the Micro-rise vee-block set shown in Figure 11.15)
3. Dial indicator that discriminates to 10% or less of the total runout tolerance
4. Surface gage (or height gage) to hold the dial indicator

STEP 1 Carefully rotate and slide the equipment onto the surface plate.

STEP 2 If using a matched vee-block set, place at least three equal gage block stacks under one of the vee blocks to account for the offset of the different diameters. If using the Micro-rise vee-block set (shown in Figure 11.15) adjust one of the vee blocks directly to account for the offset in the two datum

180 Chapter 11 Circular Runout

Figure 11.15 Micro-rise precision adjustable vee-block set. Courtesy of Custanite Corp.

diameters and then lock it in place. (*Note:* In this case, the entire length of one of the datum diameters has not been contacted due to the small part being used. In practice, the datum should be fully contacted.)

STEP 3 Mount the part in the matched offset vee blocks so that datums *B* and *C* are contacted in their entire length, as shown in Figure 11.16.

Figure 11.16 The part is mounted on the offset datum diameters and is ready for inspection. Courtesy of Custanite Corp.

STEP 4 Place the dial indicator at top dead center on one end of the controlled diameter surface, rotate the part 360°, and find the FIM value, as shown in Figure 11.17. Repeat this step in at least three locations along the shaft (end, middle, opposite end). Each FIM value should be equal to or less than the drawing runout tolerance.

Runout tolerances are very useful in designing products, specifically products with rotating parts. The variety of circular runout applications leads the observer to some rather interesting setups, but most of them have been covered by the methods previously discussed in this chapter.

Figure 11.17 FIM values are found at different circular elements of the controlled diameter. Courtesy of Custanite Corp.

REVIEW QUESTIONS

1. Runout tolerances primarily apply to rotating parts. *True* or *False?*
2. Circular runout provides inherent control of which of the following: (a) flatness, (b) concentricity, (c) circularity, (d) both b and c?
3. Circular runout must be inspected at various _____ elements of the controlled diameter.
4. Circular runout is always RFS. *True* or *False?*
5. Circular runout must always be specified to a least one datum. *True* or *False?*
6. Circular runout *cannot* be applied to a cone. *True* or *False?*

7. In cases where circular runout has been applied to an end face of a shaft, the following is inherently controlled by the runout tolerance: (a) perpendicularity, (b) circularity, (c) cylindricity, (d) location.
8. Circular runout and concentricity tolerance mean the same thing. *True* or *False?*
9. Circular runout automatically controls roundness at each circular section of the controlled diameter. *True* or *False?*
10. To measure circular runout, the part must be constantly rotated to obtain the FIM value. *True* or *False?*

12

Total Runout

OBJECTIVES

Upon completion of this chapter, the reader should be able to:

1. Inspect total runout on a diameter or an end surface.
2. Understand the other geometric controls inherently provided by total runout.
3. Further understand compound datums.
4. Explain partial datums.
5. Explain how total runout controls taper.

INTRODUCTION AND APPLICATIONS

Total runout is a composite control for rotating parts that is always specified with respect to at least one datum. Total runout applied to the diameter of a rotating part inherently controls circularity, concentricity, straightness, taper, and part surface profile. Total runout applied to an end surface of a rotating part inherently controls wobble, perpendicularity, and flatness of the surface. Total runout is typically applied to control mass in high-speed rotating parts for the purpose of balance, vibration, and other mass-related problems. Total runout differs from circular runout primarily because total runout is evaluated in terms of one collective full indicator movement (FIM), whereas circular runout applies to each circular element of the diameter (several different FIMs).

The measurement of total runout involves one primary difference in the methods used for circular runout in Chapter 11. As covered in Chapter 11, circular runout requires various FIM values obtained at different circular elements of the controlled feature. Total runout, however, is the combination of *all circular elements viewed collectively*. Therefore, when measuring total runout, the dial indicator is traversed in a straight line on the controlled feature to obtain *one composite FIM*. The dial indicator tip never leaves the part during the measurement.

KEY FACTS

1. Total runout must be specified with respect to at least one datum reference at all times. The datum feature establishes a datum axis of rotation.
2. The tolerance zone for total runout is (technically) *two lines* where the surface of all circular elements must lie. The measurement of total runout is *one* FIM value of all circular elements simultaneously while the part is rotating about the datum axis 360°.
3. Total runout inherently controls circularity, concentricity, straightness of surface elements, and taper, because the cumulative variations of these geometric errors are all seen in this composite surface-to-axis relationship.
4. Runout tolerances are always regardless of feature size (RFS).
5. Total runout, when applied to the end surface of a part, controls perpendicularity, wobble, and flatness by virtue of the composite FIM value.

TOTAL RUNOUT: OUTSIDE-DIAMETER DATUM

Consider the total runout requirement shown in Figure 12.1. The feature control frame states that the surface of the controlled diameter must be within a total runout tolerance of .002″ (FIM is understood) with respect to datum axis A. The tolerance zone for this total runout requirement is *two perfectly straight and parallel lines that are .002″ apart where the entire surface of the controlled diameter must lie,* as shown in Figure 12.2. This requires that the controlled diameter must not exceed a total runout of .002″ with respect to datum A. Therefore, datum A must be contacted and the part rotated 360°, and a dial indicator must be moved in a straight line along the axis of the controlled diameter.

The following equipment is required for this inspection method:

1. Surface plate
2. Dial indicator that discriminates to no more than 10% of the total runout tolerance
3. Vee block large enough to contain datum diameter A
4. Surface gage (or height gage) to hold the dial indicator
5. Precision parallel (or other precision block) to guide the base of the surface gage or height gage in a straight line parallel to the axis of the controlled diameter

STEP 1 Carefully rotate and slide the vee block and surface gage onto the surface plate.

STEP 2 Mount the part (on datum A) in the vee block.

STEP 3 Bring the precision parallel (or other precision block) in contact with the side of the vee block (if necessary) so that the base of the surface gage can be guided by the parallel during the measurement. (*Note:* At times, the base

Total Runout: Outside-Diameter Datum

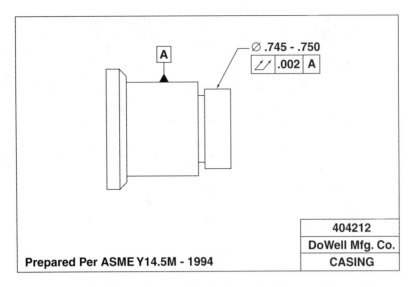

Figure 12.1 Drawing requirement for total runout with respect to one datum feature.

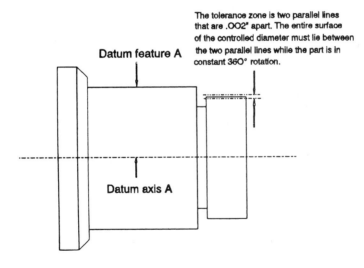

Figure 12.2 Tolerance zone for the part.

of the surface gage can be guided by the edge of the vee block that is parallel to the axis of the part in the vee block.)

STEP 4 Position the dial indicator at top dead center of one end of the controlled diameter. Make sure that there is sufficient travel on the indicator for the measurement.

186 Chapter 12 Total Runout

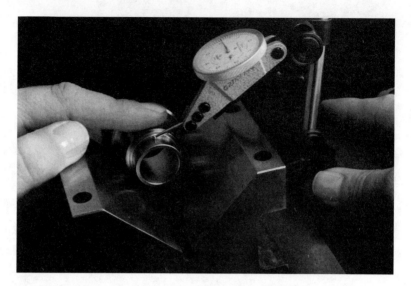

Figure 12.3 The part has been set up for total runout measurement.

STEP 5 While rotating the part in a constant 360° motion, guide the dial indicator along the top of the controlled diameter until it reaches the opposite end of the diameter. Watch for the FIM value that results during the traversing of the indicator. The setup and method are shown in Figure 12.3.

If the composite FIM value is equal to or less than the total runout tolerance, the part is acceptable. If the composite FIM value exceeds the total runout tolerance, the part is not acceptable.

As the observer noticed during these steps for total runout measurement, the dial indicator tip, in effect, spiraled the part while it was rotating in the vee block. This is the reason that total runout tolerance inherently controls several geometrical errors that may exist.

TOTAL RUNOUT: INSIDE-DIAMETER DATUM

At times, total runout requirements call for inside-diameter datums such as the drawing requirement in Figure 12.4. In this case, an inside diameter establishes the axis of rotation for total runout measurement. The drawing in Figure 12.4 requires that the largest outside diameter of the part shall not run out from datum diameter *A* more than .003″ total runout. In this case, contact of the datum diameter should be made with the largest true gage pin that can be inserted into the diameter. There are some choices that could be made here, such as a gage pin (for use in a vee block) or a gage pin with precision centers (for use on adjustable centers). Either of these will make proper contact of the inside-diameter datum and establish a functional axis of rotation for the total runout measurement.

Total Runout: Inside-Diameter Datum   **187**

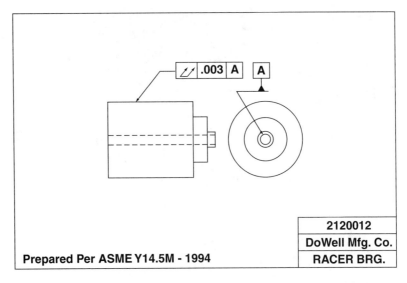

Figure 12.4 Drawing requirement for total runout with respect to an inside-diameter datum.

The following equipment is required for this method:

1. Surface plate
2. Dial indicator that discriminates to no more than 10% of the total runout tolerance
3. Largest gage pin that can be inserted into datum *A* diameter
4. Matched vee blocks (or adjustable centers if the gage pin has centers) (*Note:* The following method uses an adjustable vee-block set with a base.)
5. Precision parallel (or other precision block) to guide the surface gage (if necessary)

STEP 1   Carefully slide the vee blocks and surface gage onto the surface plate.

STEP 2   Insert the gage pin selected into the inside-diameter datum *A* in a manner such that the part will be between the two vee blocks in the setup.

STEP 3   Mount the exposed sections of the gage pin into the matched vee blocks, as shown in Figure 12.5.

STEP 4   The options at this step are (1) to set the vee blocks apart in a manner such that the observer can rotate and slide the gage pin back and forth during the measurement without moving the base of the surface gage or (2) to mount the part between the vee blocks and ask someone's help to constantly rotate the part in the vee blocks while the observer moves the surface gage in a straight line across the top dead center of the controlled diameter. In either case, the part must be constantly rotated while the indicator is moved in a straight line to obtain one composite FIM value for total runout. In the

188 Chapter 12 Total Runout

Figure 12.5 The part is set up for total runout measurement. The dial indicator is set at top dead center on one end of the controlled diameter.

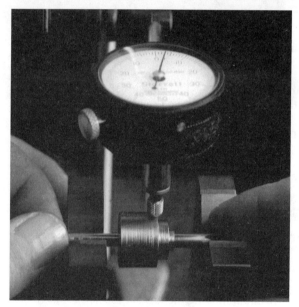

Figure 12.6 The dial indicator has been traversed in a straight line to the opposite end of the controlled diameter (without leaving the part) to obtain a composite FIM value.

method shown in Figure 12.5 and concluded in Figure 12.6, the observer can rotate and slide the part at the same time to obtain the FIM value. As before, the part is acceptable if the composite FIM is equal to or less than the total runout tolerance.

Using this method, the gage pin was inserted into the datum hole and provided a functional axis of rotation for the runout measurement. The observer was able to measure total runout without assistance since the surface gage holding the dial indicator remained stationary during the measurement.

TOTAL RUNOUT: END SURFACE

Some design applications require circular or total runout on an end surface with respect to a datum axis. Total runout applied to an end surface inherently controls wobble, perpendicularity, and flatness of the surface to the extent of the total runout tolerance. Refer to the drawing in Figure 12.7. The feature control frame states that the end surface of the part must be within a total runout tolerance of .002" (a composite FIM is under-

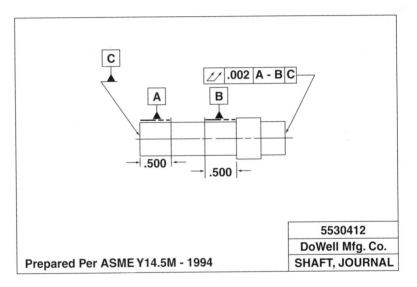

Figure 12.7 Drawing requirement for total runout on an end surface.

stood) with respect to primary multiple datums *A* and *B* and secondary datum *C*. The tolerance zone for this requirement is *two perfectly parallel planes that are .002" apart*, as shown in Figure 12.8. The entire end surface must lie between these tolerance zone planes during constant 360° rotation of the part on its datums.

This drawing requires that the end surface of the part must be within a total runout of .002" with respect to the datum axis. The datum axis is established by compound *primary datums A* and *B* and *secondary* alignment datum *C*. Note, in this case, that the primary multiple datum is actually two different lengths of the same diameter on the part. This example shows how a *chain line* can be used to indicate *partial datums*. A partial datum is a specific length or area of a datum feature that establishes the datum.

190 Chapter 12 Total Runout

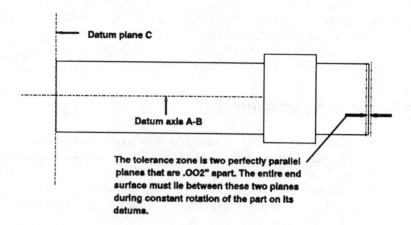

Figure 12.8 Tolerance zone for the part.

Compound datum features (refer to Chapter 2) are two different features that are used together to establish a datum.

Primary datums A and B must be contacted by a simulated datum feature that covers a .500″ (basic dimension) length of each part of the diameter. For inspection purposes, multiple datums A and B are established first; then the secondary datum feature C is contacted with a plane at 90° from the primary datum. The following equipment is required for this method:

1. Surface plate
2. Dial indicator that discriminates to no more than 10% of the total runout tolerance
3. Surface gage (or height gage) to hold the dial indicator
4. Pair of matched vee blocks (the proper set would be vee blocks that are .500″ wide)
5. Precision parallel (or other precision block) to contact the secondary datum C
6. Right-angle plate (knee) to guide the surface gage in a straight line perpendicular to the datum axis

STEP 1 Carefully slide the vee blocks, surface gage, right-angle plate, and parallel onto the surface plate.

STEP 2 Mount the part in the matched vee blocks on partial datums A and B, with datum A in one vee block and datum B in the other.

STEP 3 Bring the precision parallel into a position that is 90° from the axis derived by the vee blocks, and bring datum C of the part against the parallel.

STEP 4 Bring the dial indicator tip in contact with the center of the controlled surface, as shown in Figure 12.9. Establish enough travel on the indicator for the measurement.

Figure 12.9 The part has been set up for end-surface total runout inspection.

STEP 5 Position the right-angle plate (knee) at 90° from the edge of the vee blocks; then bring it into contact with the base of the surface gage and the edge of the vee blocks, as shown in Figure 12.9.

STEP 6 While rotating the part in the vee blocks (and against the parallel), move the dial indicator in a straight line (guided by the right-angle plate) until the indicator tip is at the outer edge of the controlled face, as shown in Figure 12.10. During this movement, watch for the composite FIM value. If the FIM is equal to or less than the total runout tolerance, the part is acceptable.

As we have seen in this chapter, total runout is a composite control over cylindrical and flat surfaces with respect to a datum axis, and yet it is not a control that is overly difficult to measure. The important thing to remember is to keep the dial indicator or probe traversing in a straight line during the measurement and to look for a composite FIM value, not individual FIM values.

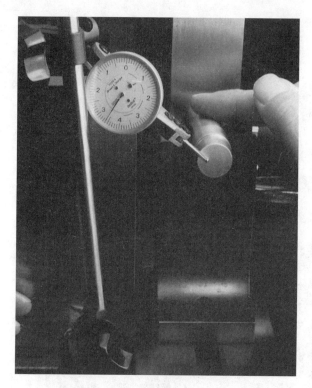

Figure 12.10 The indicator is moved in a straight line outward while the part is rotated 360°.

REVIEW QUESTIONS

1. Datums are not required with total runout. *True* or *False?*
2. The FIM value for total runout is a composite value for the entire surface. *True* or *False?*
3. Total runout, when applied to a diameter, inherently controls the following: (a) taper, (b) concentricity, (c) circularity, (d) all of these, (e) none of these.
4. Total runout, when applied to an end surface, inherently controls the following: (a) flatness, (b) perpendicularity, (c) wobble, (d) all of these, (e) none of these.
5. When inspecting total runout at the surface of a diameter, the dial indicator must traverse in a straight line across the surface without leaving the part. *True* or *False?*
6. Total runout cannot be applied to a cone. *True* or *False?*
7. Which of the following is required for total runout inspection?
   (a) A composite FIM value
   (b) At least a primary datum reference
   (c) Constant rotation of the part during measurement
   (d) All of these
8. The tolerance zone for total runout at the surface of a diameter is (a) a cylinder, (b) two parallel lines, (c) two parallel planes, (d) none of these.
9. The chain line is sometimes used to limit the control of a geometric tolerance. In this chapter, the chain line was used on the datum for total runout to show that it was a _____ datum.
10. Total runout tolerance is always considered at RFS. *True* or *False?*

13

Profile of a Line

OBJECTIVES

Upon completion of this chapter, the reader should be able to:

1. Explain the difference between profile controls that use datums and those which do not.
2. Explain when profile tolerance zones are bilateral and when they are not.
3. Understand why basic dimensions are required to define the true profile.
4. Use an optical comparator for profile measurement.
5. Understand the use of mechanical gaging for profile measurement.

INTRODUCTION AND APPLICATIONS

Profile tolerances are often used to control irregular shapes (for example, contours). Profile tolerances in general are the only family of geometric tolerances for which the designer has the option to specify a datum reference or not. In general, if datums are not specified for a profile tolerance, the tolerance provides control over *shape* only. When datums are used, the profile tolerance collectively controls the *shape* of the contour and the *size and/or location* of the contour in one tolerance zone. A typical application for profile tolerances is mating part contours.

Profile tolerance zones take the shape of the basic profile (defined by basic dimensions) and are bilateral around that profile unless otherwise specified. (See the key facts that follow for more information.)

KEY FACTS

1. Profile tolerances are the only geometric tolerancing symbols that *sometimes* use datums.
2. Profile tolerances are applied to parts with irregular shapes, such as contours.
3. Profile tolerances require that the irregular-shaped surface must be defined using basic dimensions to define the true profile.
4. The tolerance zones for profile automatically take the shape of the surface being controlled and surround the basic profile bilaterally (unless otherwise indicated).
5. Unilateral profile tolerances are indicated using a phantom line to show the direction and amount of the tolerance zone from the basic profile.
6. The tolerance zone for the profile of a line is two lines that equally surround the basic profile and are the stated tolerance apart. These lines equally surround the basic profile, that is, unless phantom lines are used to show unilateral direction (as previously stated).
7. Profile tolerances, when no datum is specified, control the shape of the surface profile.
8. Profile tolerances, when a datum is specified, control the shape, size, and/or location of the surface profile.
9. Arcs, radii, curved lines, surfaces, or lines (all basic) may be used to describe the true profile.
10. All profile measurements, especially on contoured surfaces, must be made with the probe (or indicator) at 90° from the tangent line of the surface.
11. Maximum material condition (MMC) or regardless of feature size (RFS) principles are usually not applicable to profile tolerances (except for size datums).
12. Profile of a surface tolerance can be applied to interrupted surfaces to control coplanarity of these surfaces.

GENERAL MEASUREMENT OF PROFILE TOLERANCES

A limited number of methods and equipment can be used to measure profile tolerances due to the nature of the tolerance and the irregularities of the parts where profile is used. Profile tolerances are typically applied to parts with irregular shapes such that standard inspection equipment will often not suffice. There are, however, methods and equipment that can be used to measure profile tolerances such as optical comparators (with appropriate overlays), coordinate measuring machines, limit gages, and profile gage designs (or hard tooling).

The inspection method and equipment depend largely on the size and configuration of the part and the datum structure (when datums are used). Note, at this point, that the use of datums for profile tolerances is optional, depending on design requirements. Profile tolerances are the only types of tolerances in geometric tolerancing that use datums or not, depending on the functional requirement. Other geometric tolerances either always or never reference datums.

PROFILE DIMENSIONING AND TOLERANCE ZONES

The tolerance zones for profile tolerances depend largely on the shape of the surface being controlled and the way the profile tolerance is shown on the drawing. Unless otherwise specified, profile tolerances are always bilateral around the true profile. The true profile of irregular shapes must always be defined by basic dimensions; then the profile tolerance zone surrounds the basic profile. Refer to Figure 13.1 for examples of profile tolerance zones.

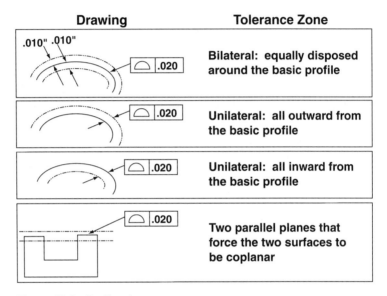

Figure 13.1 Profile tolerance zones.

Notice in Figure 13.1 that the tolerance zones are equally disposed about the true profile unless *phantom lines* are used to indicate unequal tolerances. Unequal tolerances, always shown by phantom lines, can be stated from the basic profile *all outward, all inward,* or a different combination, such as *some amount outward and a different amount inward.* Also notice a special application for profile of a surface tolerance that controls *coplanar* surfaces or forces them to be in the same plane within tolerances.

PROFILE OF A LINE: NO DATUMS

When collective control over the entire contoured surface is not necessary, and line element control will suffice for function, a profile of a line tolerance will be applied to the contour. Profile of a line, as shown in Figure 13.2, provides a tolerance zone of *two profile boundaries the stated tolerance apart and shaped like the basic profile,* as

196   Chapter 13   Profile of a Line

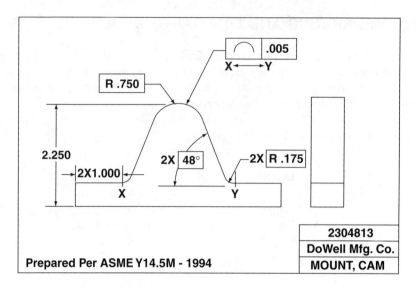

Figure 13.2   Drawing requirement for profile of a line.

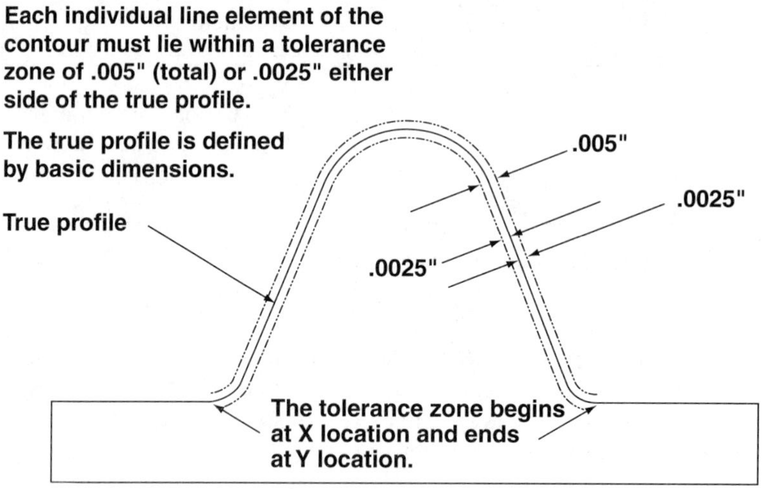

Figure 13.3   Tolerance zone for the part.

shown in Figure 13.3. This tolerance zone, when datums are not specified, controls the shape of individual line elements of the contoured surface.

Each line element of the controlled surface (individually) must lie within the profile tolerance zone. When datums are not specified, the profile tolerance controls only the shape of each line element of the contour. Location or size must be separately dimensioned and controlled.

OPTICAL COMPARATOR MEASUREMENT

The most widely used equipment for measuring profile tolerances on contours is the optical comparator and an overlay chart. The overlay chart tolerance lines, drawn to the basic profile and scaled to match the magnification power of the comparator (for example, 10×, 20×, 30×, 50×, etc.), can be constructed with two tolerance lines that represent the profile tolerance zone. These lines would be constructed bilaterally from the basic profile unless the drawing indicated unilateral tolerances (using phantom lines). The overlay chart is mounted to the comparator screen, and the contoured surface being controlled is placed in front of the light source of the comparator.

In this case, since datums have not been specified, the part does not have to remain stationary on the comparator stage for the measurement. The image of the contoured surface of the part can be rotated or moved (or the overlay can be moved) to see if the line element being viewed will lie between the tolerance lines on the overlay chart. The act of moving the part (or the overlay) is called "best fit". One drawback of visual methods on an optical comparator is that the comparator sees only the outline of the part.

Due to this limitation, visual methods work well only with parts that have thin cross sections (for example, gaskets, washers, and plate stock). For thicker parts with contours, accessories to the comparator are usually necessary, such as pantograph mechanisms with a stylus for tracing the line elements of the part contour and a reticle follower that can be seen on the screen with respect to the overlay tolerance lines. The stylus can then be used to trace the line elements of thicker parts (one at a time) to see if they fall within the boundaries of the profile tolerance zone. The overlay and the reticle follower are used by the observer to evaluate the part for visual inspection between the tolerance lines. The observer must keep in mind that when no datums are specified, the part can be moved or rotated for best fit between the tolerance lines. When datum features have been specified for a profile tolerance, the comparator is equipped with a stage that moves in X (left and right), Y (up and down), and Z (in and out) directions to measure the amount of deviation of the part. Overlay lines are basically attribute (go/no-go) inspection, and a movable stage is used for variable measurements.

An example of the part being inspected to overlay lines on an optical comparator is shown in Figure 13.5. In this method, the part, remember, does not have to remain stationary on the comparator stage. If a portion of any line element falls outside the overlay chart tolerance lines, move or rotate the part to see if the entire line element can be *best-fitted* into the tolerance lines. If the element can be best-fitted to where it falls anywhere between the tolerance lines, the line element is acceptable.

When a profile tolerance is out of the tolerance lines on the overlay, the optical comparator stage (equipped with liner scales) can measure the deviations directly. For example, the part in Figure 13.6 has a requirement for the shape and size of the tip to be controlled by a profile tolerance. If the radius is acceptable (per the overlay), but the height and width are out of tolerance, the comparator stage can be used to measure the error. Figure 13.7a and b show the optical comparator stage being used to measure the error in height, and Figure 13.7c and d show the comparator stage being used to measure the error in width.

Figure 13.4 An optical comparator. Courtesy of Barry Controls Aerospace, Burbank, CA.

Figure 13.5 The part profile tolerance is being inspected on an optical comparator using overlay tolerance lines. Courtesy of MTI Corporation.

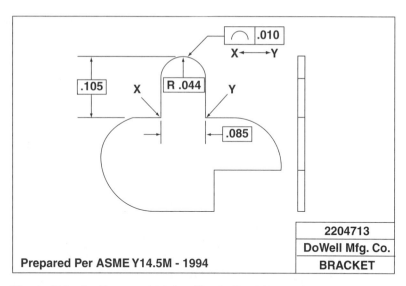

Figure 13.6 Another example of profile of a line tolerance.

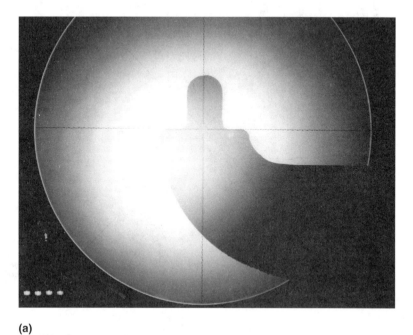

(a)

Figure 13.7 (a) and (b) Measuring height error on a profile tolerance using the comparator stage. Courtesy of Gage Master Corporation.

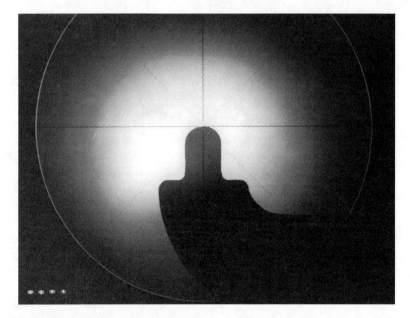

(b)

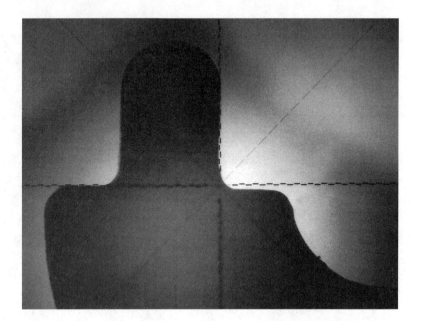

(c)

Figure 13.7 (c) and (d) Measuring width error on a profile tolerance using the comparator stage. Courtesy of Gage Master Corporation.

(d)

Figure 13.7 *(continued)*

PROFILE OF A LINE USING DATUMS

An example of the same part previously discussed is shown in Figure 13.8, except that in this example the drawing has been changed to specify datum references for the profile tolerance. The tolerance zone for this requirement is *two lines that are .005" apart around the basic (or true) profile and located by the basic dimensions from the datum references,* as shown in Figure 13.9.

The tolerance zone is bilateral (plus or minus .0025" from true profile) in this case. When datum references are specified, the profile tolerance controls more than just the shape of the contour. It controls shape and location simultaneously. In this case, as we see in the drawing, other basic dimensions are needed to locate the profile tolerance zone from the datum reference. Since these basic dimensions are shown from datum references, the profile tolerance zone is fixed in space. This means that, once the part has been referenced for measurement and moved into the overlay tolerance lines, it cannot be moved or rotated further for best fit. The line element of the contour must fit when the part is compared to the tolerance lines. An example of this comparator measurement is shown in Figure 13.10.

Chapter 13 Profile of a Line

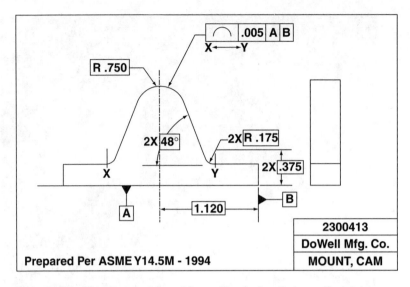

Figure 13.8 Drawing requirement for profile of a line that specified datums.

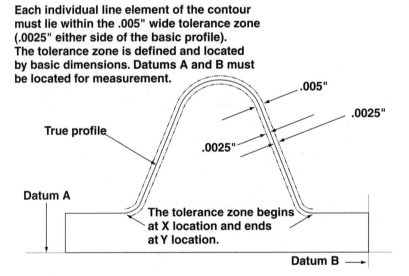

Figure 13.9 Tolerance zone for the part.

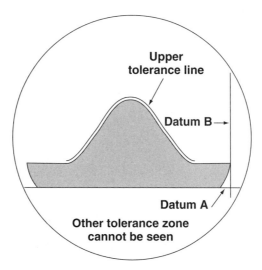

Figure 13.10 Optical comparator with overlay (datums applicable). The part is indexed on the comparator stage and moved into the tolerance zone lines. No movement is allowed since datums are specified.

MECHANICAL GAGING: PROFILE TOLERANCES

In many cases, a profile tolerance gage can be designed and constructed when that is the appropriate technical and economical decision for inspection. A sample of a profile gage is shown in Figure 13.11. In this case, the gage has been designed to inspect the part in Figure 13.8 with respect to the primary and secondary datums (*A* and *B*, respectively). As shown in Figure 13.11, the gage has been designed with a base plate and a side plate to locate the primary datum on its three or more highest points and the secondary datum on its two or more highest points. A dial indicator is mounted on a guide bar that follows the master contour in a manner such that the dial indicator is always positioned at 90° from the tangent line of the part contour. In this manner, the dial indicator is mastered at zero (where zero equals the basic contour required), and then the part is evaluated with respect to plus and minus values along the line element that is being traced. Since the requirement is profile of a line, each individual line element that is traced must fall within the bilateral tolerance zone for profile. The dial indicator is positioned in a manner such that it can be moved from side to side, allowing for different line elements to be traced by the observer.

204   Chapter 13   Profile of a Line

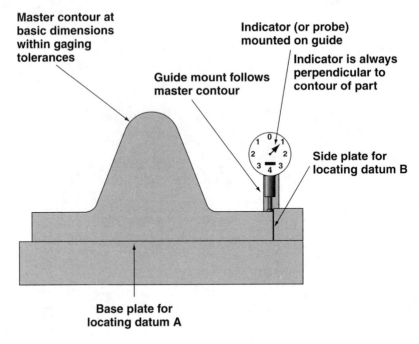

Figure 13.11   Example of a variable gaging fixture for the profile tolerance.

PROFILE: TO LOCATE A SURFACE

A new application in the 1994 ASME standard is the use of profile tolerance to locate a surface, as shown in Figure 13.12. The drawing shows a pin that is perpendicular to datum A on the part; however, the profile of a line symbol has been applied. One important thing to remember in this drawing is that, whenever features are shown at right angles, a 90° angle is understood. In this case, the feature is shown at a right angle to datum A and the profile tolerance causes the angle to be interpreted as 90° basic. Profile of a line controls surface line elements of the pin. In this case, each surface line element of the pin must be perpendicular within a tolerance zone of two parallel lines that are .0002″ apart and stand 90° to datum A.

The tolerance zone is shown in Figure 13.13. The designer, in this case, could have used perpendicularity to control the axis of the pin, but chose to use profile instead. The reason profile was used instead of perpendicularity is that perpendicularity applies to the axis and creates a virtual condition beyond MMC size. The profile tolerance, in this case, cannot create a virtual condition *and* controls the perpendicularity and straightness of the surface at the same time. It should be noted, however, that in many cases, this profile tolerance would be more control than necessary.

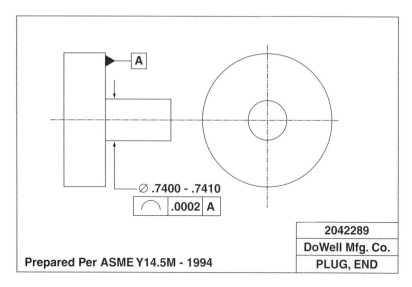

Figure 13.12 Profile of a line to locate a surface.

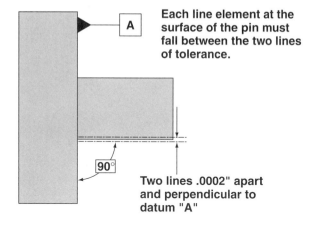

Figure 13.13 The profile tolerance zone.

REVIEW QUESTIONS

1. Measurements of profile tolerances must be made with the indicator (or probe) at 90° to the _____ line of the contour.
2. Of all the geometric tolerances, profile tolerances are the only family for which datums can be used or not, depending on functional requirements. Other tolerances either never use datums or must have datums specified. *True* or *False?*

3. Profile of a line controls:
   (a) The entire contoured surface simultaneously
   (b) Each line element of the contoured surface
   (c) The shape and size at all times
   (d) None of these
4. Profile tolerances require that the true profile is described by basic dimensions. *True* or *False?*
5. All profile tolerance zones are bilateral from the basic profile unless otherwise specified. *True* or *False?*
6. Profile tolerances normally apply to parts with irregular shapes (such as contours). *True* or *False?*
7. When datums have not been specified for a profile tolerance applied to a contour, which of the following statements is true?
   (a) Shape, size, and/or location are controlled
   (b) Size control is all that is gained
   (c) Only the shape of the basic contour is controlled
   (d) None of these
8. Where unilateral profile tolerances are required by design, the _____ line can be used to show the direction and amount of tolerance.
9. Which of the following inspection equipment is most often used to inspect profile tolerances: (a) coordinate measuring machine, (b) optical comparators, (c) sweep gages, (d) surface-plate setups?
10. When datums are specified for a profile tolerance of a contour, basic dimensions must not only define the contour, but also locate it from the specified datums. *True* or *False?*

14

Profile of a Surface

OBJECTIVES

Upon completion of this chapter, the reader should be able to:

1. Understand the difference between profile of a line element and profile of a surface tolerance.
2. Measure coplanar surfaces that are controlled by profile of a surface tolerance (datums or no datums).
3. Understand how profile of a surface (all around) can be used to define tolerances for control of conical shapes.

INTRODUCTION AND APPLICATIONS

Profile of a surface tolerance is a composite control over irregular shapes (such as contours previously discussed in Chapter 13). The applications for profile of a surface tolerance are the same for profile of a line, but there are more of them. Some of the extended applications for profile of a surface include (1) control of two or more interrupted surfaces that must be in the same plane (often called *coplanarity*) and (2) control of a cone to ensure a true cone within tolerances (often called *conicity*).

Profile of a surface *coplanarity* application examples are datum target pads on a casting or compound datum surfaces on a machined part in order to *qualify* the datum. *Conicity* applications include mating conical sections of parts, such as expanding mandrels, shutoff valves, and other similar applications.

KEY FACTS

1. Profile tolerances are the only geometric tolerancing symbols that *sometimes* use datums.
2. Profile tolerances are applied to parts with irregular shapes (such as contours).
3. Profile tolerances require that the irregular-shaped surface be defined using basic dimensions to define the true profile.
4. The tolerance zones for profile automatically take the shape of the surface being controlled and surround the basic profile bilaterally (unless otherwise indicated).
5. Unilateral profile tolerances are indicated using a phantom line to show the direction and amount of the tolerance zone from the basic profile.
6. The tolerance zone for profile of a surface is three-dimensional and extends along the length and width (or circumference) of the part feature being controlled. The tolerance zone equally surrounds the basic profile, that is, unless phantom lines are used to show unilateral direction (as previously stated).
7. Profile tolerances, when no datum is specified, control the shape of the part profile.
8. Profile tolerances, when a datum is specified, control the shape, size, and/or location of the part profile.
9. Arcs, radii, curved lines, surfaces, or lines (all basic) may be used to describe the true profile.
10. All profile measurements, especially on contoured surfaces, must be made with the probe (or indicator) at 90° from the tangent line of the surface.
11. Profile of a surface is the only geometric tolerance (other than size tolerances per rule 1) that controls the *coplanarity* of interrupted surfaces.
12. Profile of a surface can also be used to control a conical part.
13. Maximum material condition (MMC) or regardless of feature size (RFS) principles are usually not applicable when using profile tolerances (except for size datums).

PROFILE OF A SURFACE: NO DATUMS

Profile of a surface is an extended control compared to profile of a line tolerance because all line elements of the part contour must lie within one envelope of tolerance. To demonstrate the difference, the profile of a line part from Chapter 13 is shown in Figure 14.1, but it has been changed from profile of a line to profile of a surface tolerance. The tolerance zone, in this case, is *two imaginary contoured surfaces that surround the basic (or true) profile and are .005" aparti as shown in Figure 14.2.*

The same inspection methods covered in Chapter 13 can be used on this part except that the line elements that are traced by the observer must all be related to each other within an envelope of tolerance, instead of each line element being measured and controlled separately. To do so, the observer must relate high and low areas of one line element to high and low areas of another element.

For example, a part will be used that is acceptable to the profile of a line tolerance zone, but would be rejected when compared to the profile of a surface tolerance zone.

Profile of a Surface: No Datums   **209**

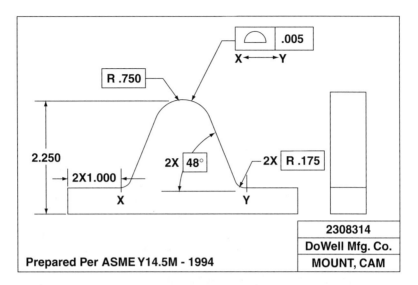

Figure 14.1 Drawing requirement for profile of a surface tolerance (no datums).

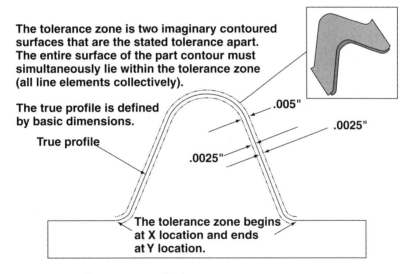

Figure 14.2 Tolerance zone for the part.

The bilateral profile tolerance for the example part is the basic profile shape plus or minus .010″ (or .020″ tolerance zone), and two line elements are traced by the observer. Line element 1 is traced by the observer and has a low area of −.012″ and a high area of +.005″. Since the part requirement has no datums specified, the part can be shifted or rotated to fit the tolerance zone (the total deviation is .017″). Line element 2 is traced by the observer and has a low area of $-.002''$ and a high area of $+.017''$. Again, if the tolerance applicable were profile of a line tolerance with no datums, the part could be shifted or rotated to fit the tolerance zone (since the total deviation of the line element is .019'').

However, if the part profile tolerance were profile of a surface, the worst areas must be evaluated simultaneously ($-.012''$ on line element 1 and $+.017''$ on line element 2). Shifting or rotating the part will not be successful since the combined deviation of .029'' exceeds the total tolerance band of .020''.

The inspection of profile of a surface on contours and irregular shapes must take into account the entire surface with respect to the *envelope* of tolerance required. Since no datums are specified on the part, the part can still be shifted or rotated for best fit, but only when the extremes of different line elements allow for acceptance of the part after the shift or rotation has been made.

The previous example part would not be accepted simply because the combined error of different line elements exceeded the .020'' tolerance zone. If the combined error of the line elements of the part were less than the total tolerance zone, shifting or rotating the part should prove successful.

PROFILE OF A SURFACE USING DATUMS

An example drawing for profile of a surface using datums is shown in Figure 14.3. This is the same part that was used in Chapter 13 for profile of a line using datums. The tolerance zone in this case consists of *two imaginary contoured surfaces that are .005''*

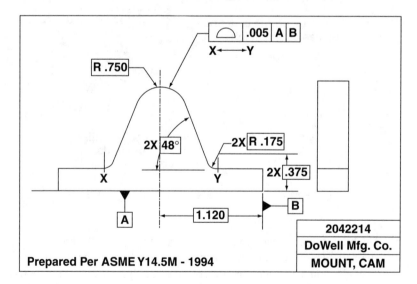

Figure 14.3 Drawing requirement for profile of a surface tolerance that specified datums.

apart and surrounding the basic (or true) profile. This zone is also located by basic dimensions from the datum references, as shown in Figure 14.4. Once again, the profile measurement methods covered in Chapter 13 (for example, optical comparator and sweep gage) can be used for profile of a surface tolerance with respect to datums references when the observer considers the differences previously covered *and* the fact that the part, in this case, *cannot be shifted or rotated* for best fit. Profile of a surface tolerance with respect to datums is one of the most difficult tolerances to produce because of its all-encompassing control on the contour or irregular shape of the part.

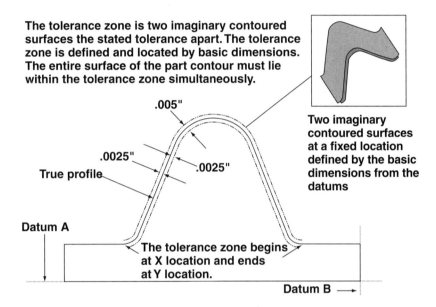

Figure 14.4 Tolerance zone for the part.

PROFILE FOR COPLANAR SURFACES: INTRODUCTION

Another design application for profile of a surface tolerance is the control of coplanarity of two or more surfaces. Once again, datums may or may not be specified. At times, coplanarity of surfaces is confused by some observers as being a parallelism or flatness inspection. Keep in mind that any interrupted surface of a part can be parallel (or flat), but not in the same plane as other surfaces. Coplanarity requires that each controlled surface must lie in the same plane as other controlled surfaces. Parallelism and flatness apply only to individual surfaces. Refer to Figure 14.5 for examples of the difference.

PROFILE FOR COPLANAR SURFACES: DATUMS APPLIED

Coplanar surface control using profile of a surface tolerance and datums is a fairly straightforward application. An example requirement for coplanarity using profile of a

212   Chapter 14   Profile of a Surface

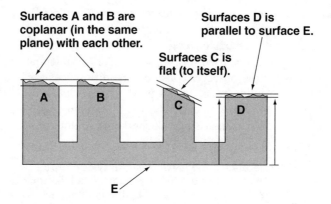

Figure 14.5   The difference between parallelism, flatness, and coplanarity.

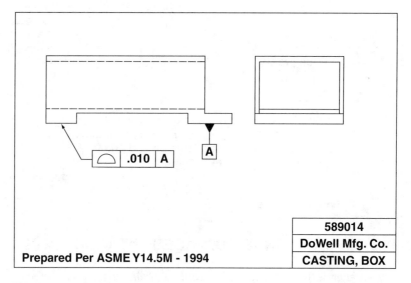

Figure 14.6   Coplanarity requirement (using profile of a surface) that specifies a datum reference.

surface tolerance (with datums applicable) is shown in Figure 14.6. The profile of a surface tolerance zone for coplanar surfaces always consists of *two parallel planes the stated tolerance apart* and *the zone equally surrounds the datum plane,* as shown in Figure 14.7.

In this case, the controlled surface must be in the same plane as the datum plane within the specified profile tolerance zone of .010″ total. The tolerance zone of .010″ is equally disposed around the datum plane (plus or minus .005″). Inspecting this part requires contacting the datum feature, setting zero on the indicator at the datum plane,

Profile for Coplanar Surfaces: Datums Applied

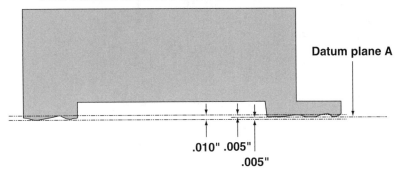

Figure 14.7 Tolerance zone for the part.

and then transferring the indicator to the controlled surface(s) to see if it is higher or lower than the datum plane (within the applicable tolerance zone).

The following equipment is required to inspect the part using the *transfer method* in open setup:

1. Surface plate
2. Dial indicator that discriminates to no more than 10% of the tolerance
3. Surface gage (or height gage) to hold the dial indicator
4. Precision parallel
5. Precision-square block (or a right-angle plate)

STEP 1 Carefully rotate and slide the measuring equipment onto the surface plate.

STEP 2 Mount the precision parallels (or other chosen accessories) onto the precision block or right-angle plate in a way such that the parallel has an overhang large enough to contact the datum feature. Next, mount the datum feature on the bottom overhang surface of the parallel so that the controlled surface of the part is facing upward, as shown in Figure 14.8. In this setup, the top surface of the precision block or right-angle plate establishes the datum (or zero) plane for the measurement, and the contacting surface of the precision parallel *transfers* that plane to the datum of the part.

STEP 3 Set zero on the dial indicator on the top surface of the precision block (or right-angle plate) as shown in Figure 14.9. At this point, zero represents the datum plane.

STEP 4 Indicate the controlled surface of the part compared to the zero setting as shown in Figure 14.10. The controlled surface cannot be higher or lower than zero by more than .005" (in this case). If any area of the controlled surface exceeds these boundaries, the surface is out of profile *coplanarity* tolerance.

214 Chapter 14 Profile of a Surface

Figure 14.8 The part is mounted for inspection using a precision-square block and precision parallel.

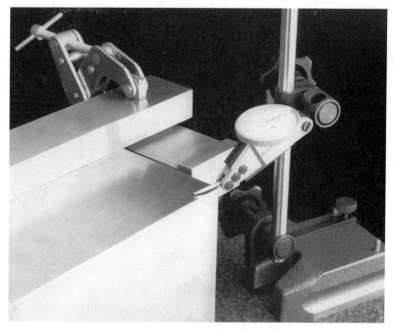

Figure 14.9 Setting zero on the dial indicator at the datum plane.

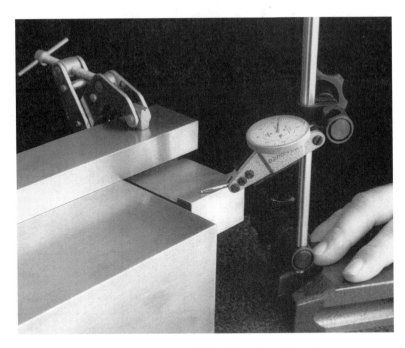

Figure 14.10 Inspecting the controlled surface for coplanarity.

PROFILE FOR COPLANAR SURFACES: DATUMS NOT APPLIED

Another example of coplanarity is one for which no datums have been specified. Figure 14.11 shows an example of a drawing that requires coplanar surfaces with an *intrinsic* datum (the surfaces to each other). The tolerance zone, in this example, consists of *two imaginary and perfectly parallel planes that are .010″ apart. These planes surround both surfaces* as shown in Figure 14.12.

The inspection method is different for this type of profile requirement from the previous example for which a primary datum feature was specified. To inspect this part, the observer must relate the surfaces *to each other* to prove whether or not the surfaces lie within the tolerance zone. An individual surface cannot be contacted (as if it were a datum) due to the possibility of measurement error because the surface may be inclined at an angle.

QUICK-CHECK METHOD: HEIGHT DIFFERENCES

One method of establishing coplanarity of the surfaces is *height measurement on the surface plate (Note:* **Caution must be taken to use this method only as a quick check to make the acceptance decision, not rejection decisions,** due to other errors that may be present in the part during the measurement.) The error of this method is such that acceptance decisions will be correct, but rejection decisions could be false. For example,

216 Chapter 14 Profile of a Surface

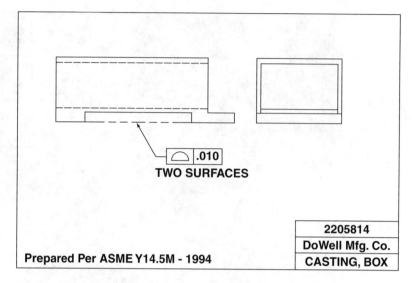

Figure 14.11 Drawing requirement (using profile of a surface tolerance) for coplanarity when no datums have been specified.

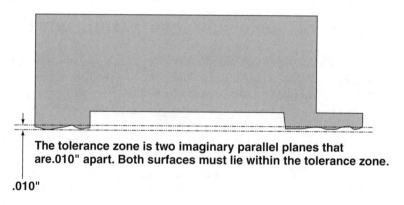

Figure 14.12 Tolerance zone for the part.

if the height measurements fall within the profile tolerance, the part is not only coplanar within the tolerance, but it is also parallel to the reference surface. If the measurements fall outside the profile tolerance, the part surfaces may be coplanar to each other, but *could be out of parallel to the reference surface* Since the reference surface is not a datum, parallelism error cannot be used to reject the part.

The following equipment is required for this method:

1. Surface plate
2. Dial indicator that discriminates to 10% or less of the profile tolerance
3. Surface gage (or height gage) to hold the dial indicator

STEP 1 Carefully rotate and slide the part and surface gage onto the surface plate.

STEP 2 Mount the part on a surface opposite the controlled surfaces (so that height measurements can be made). Keep in mind that this is simply a reference surface, not a datum feature.

STEP 3 Set zero on the dial indicator on either one of the controlled surfaces, as shown in Figure 14.13 (Note: Zero on the indicator will be used as a reference point for the measurements. Actual height measurements do not have to be made. The purpose of this method is to find the difference between the lowest and highest points of the two controlled surfaces.)

STEP 4 Keeping the zero point in mind, sweep all controlled surfaces looking for the lowest and highest point (wherever they may be). If the difference between the highest and lowest points of measurement is within the profile tolerance zone, the two surfaces are not only coplanar to each other, but also, parallel to the reference surface. In this case, the part can be accepted. If the total deviation of the lowest and highest points is equal to or less than the profile tolerance, the surfaces are coplanar within tolerance. If this value exceeds the profile tolerance, the part must be inspected more thoroughly with respect to direct coplanarity measurement.

Figure 14.13 Setting zero on the dial indicator on one of the controlled surfaces.

COPLANARITY MEASUREMENT: JACKSCREW METHOD

For direct measurement of coplanarity between two surfaces when a datum plane has not been specified, an intrinsic datum method (the surfaces measured to themselves) can be used that is very similar to the flatness inspection method covered in Chapter 4. The *jackscrew method* can be used to check the coplanar (profile) tolerance in the same manner used to check flatness of a single surface.

The only difference here is that the surfaces are interrupted. Figure 14.14 shows a sketch of the coplanarity requirement being inspected using the jackscrew method. The three jacks are placed in a triangle, and zero is set at three points using the jackscrews to raise or lower the surfaces. Once the dial indicator repeats zero set at the three jackscrew locations, the indicator is then traversed across both surfaces. If the resulting full indicator movement (FIM) value is equal to or less than the total profile of a surface tolerance (for coplanarity in this case), the part is acceptable. (Refer to Chapter 4 for more coverage on the jackscrew method.)

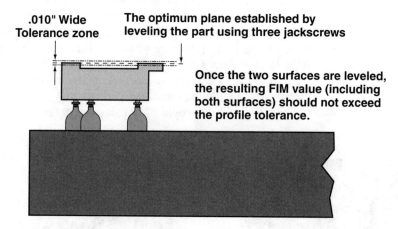

Figure 14.14 Coplanarity between the two surfaces being inspected using the jackscrew method.

COPLANARITY: COMPOUND DATUMS

Another example for controlling coplanar surfaces using profile of a surface tolerance is shown in Figure 14.15. In this case, three surfaces must be coplanar to one another and compound datums *A-B* within .005″. Datums *A* and *B* are compound primary datums because this part interfaces in assembly on both surfaces at the same time.

The tolerance zone for the control consists of two parallel planes that are .005″ apart and centered on datum plane *A-B*. Therefore, inspection of this part must prove that each of the three controlled surfaces are no higher or lower than .0025″ from plane *A-B*.

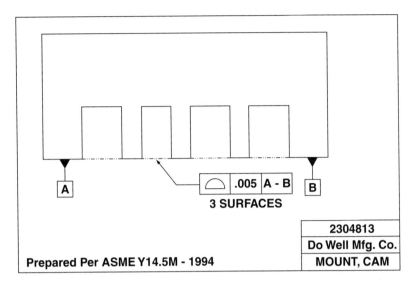

Figure 14.15 Profile to control coplanarity with respect to compound datums.

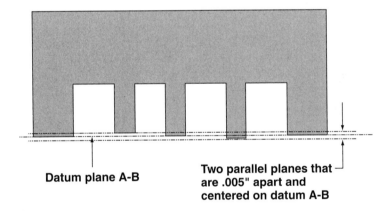

Figure 14.16 The tolerance zone for coplanarity to compound datums.

REVIEW QUESTIONS

1. Profile of a surface can be specified with or without a datum reference. *True* or *False?*
2. When the profile of a surface tolerance is applied to two surfaces in the same plane (for coplanarity control), the datums are intrinsic (to each other). *True* or *False?*
3. Profile of a surface tolerance zones are _____ dimensional.
4. Profile of a surface is the only geometric tolerance that can be used to control two or more surfaces in the same plane (coplanarity). *True* or *False?*
5. Profile of a surface tolerance, when applied to a contour, is always a bilateral tolerance zone. *True* or *False?*

6. Measurements of the profile of a contour must be made with a probe (or dial indicator) that is perpendicular to the _____ line during the measurement.
7. Most profile of a surface tolerances, when applied to contours for example, must be traced when inspecting them on an optical comparator. *True* or *False?*
8. Profile of a surface tolerance, except for coplanar applications, must always have _____ dimensions to define the true profile.
9. If profile of a surface tolerance were applied to a cone (all around), would the included angle of the cone be automatically controlled? *Yes* or *No?*
10. Profile tolerances were originally designed specifically for applications involving irregular _____.

15

Concentricity

OBJECTIVES

Upon completion of this chapter, the reader should be able to:

1. Explain the difference between concentricity and runout.
2. Understand that concentricity is not a full indicator movement (FIM) value.
3. Understand how differential measurements are used to track the axis of the controlled feature.
4. Explain how concentricity can be *verified* by measuring total runout.
5. Measure concentricity using surface-plate methods and differential measurements.

INTRODUCTION AND APPLICATIONS

Concentricity is a coaxial control that is often applied to rotating parts to control balance, mass rotation, wall thickness, and other functional aspects of products. Concentricity, however, is a very restrictive tolerance requiring detailed analysis of axial locations and, in fact, *does not* provide the proper control of mass as it was intended. It is often more economical and better for design to consider runout (for rotating parts) or position (for nonrotating parts) instead of concentricity, because runout measurements (FIM) are made on the material (mass).

Keeping in mind that runout *controls concentricity* and mass, runout is often selected by designers as a better control. The applications of runout and concentricity are nearly identical, but for nonrotating parts an option to concentricity is position tolerance, regardless of feature size (RFS). Further explanation of concentricity will help to understand its limitations and the difficulty of measurement.

KEY FACTS

1. Concentricity is an axis-to-axis relationship.
2. Concentricity tolerances always require a datum reference and that the part be rotated 360° during the measurement.
3. The tolerance zone for concentricity is a *cylinder that is perfectly concentric to the datum axis* and is the diameter specified in the feature control frame.
4. Concentricity is always an RFS application. In situations where maximum material condition (MMC) is considered, use position tolerance.
5. Concentricity requires that the observer, during measurement, locate the datum axis and the controlled axis and determine whether or not the controlled axis is within the cylindrical tolerance zone. Surface measurements (such as FIM) do not apply. Concentricity is not a FIM application. Differential measurements are usually required to inspect concentricity.
6. Applications for concentricity are rare. It is often better to use runout for rotating part coaxial control or position for nonrotating parts.

DIFFERENCE BETWEEN CONCENTRICITY AND RUNOUT

Concentricity and runout are very different geometric controls that share some common traits; therefore, they are often confused with each other. They both apply to rotating parts and are measured while the part rotates 360°. They both require a defined datum feature to establish a datum axis, and they are both used to control mass. Even though these controls are similar, the results and their measurement are significantly different.

Consider the drawings in Figure 15.1a and b. The part has a hexagonal section that has been machined on round stock. The hex section, for this example, shares the same axis as the round stock. Therefore, these two size features are *perfectly* concentric to each other since they share the same axis. Runout, however, is a *surface-to-axis rela-*

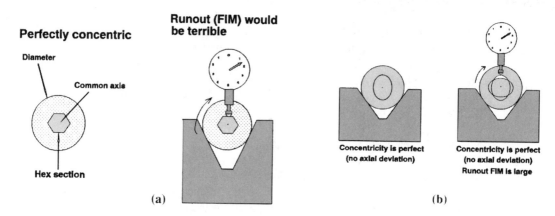

Figure 15.1 The difference between concentricity and runout: (a) a concentric hex and (b) a concentric oval.

tionship. If we were to locate this part on the diameter in a vee block and place an indicator on the surface of the hex section, the FIM result would be very bad.

This model shows a significant difference between concentricity and runout. Runout controls a surface to an axis, and concentricity controls an axis to an axis. It is for this reason that runout can be a much more efficient control for rotating parts than concentricity, *and* it is much simpler to measure.

DIFFERENTIAL MEASUREMENTS

To measure concentricity (or, better stated, *eccentricity*), we must evaluate the location of the axis of the controlled feature since concentricity applies to the axis, not the surface. Measuring axial movement is difficult, at best, with one dial indicator (or probe) due to surface problems such as lobes of circularity error, taper, and axial crookedness.

Differential measurements are required for measuring concentricity. The following example may help to understand the use of differential measurements. Since the axis is the center of two sides of the controlled feature, two opposing indicators are used to track axial location. When two indicators are used, the real axial movement is easier to measure. However, when using two indicators, the observer must understand differential measurement in order to evaluate the results.

Consider the example part in Figure 15.2, where the part is out of round and two opposing indicators are used. Notice that the first position is where the observer sets

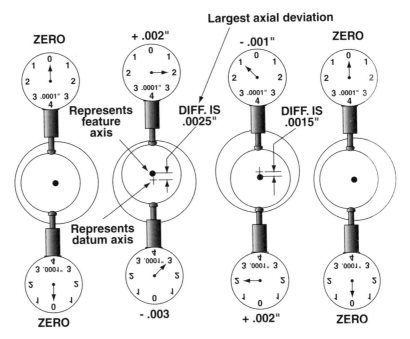

Figure 15.2 Example of two opposing indicators being used on a feature that must be concentric to a datum axis.

both opposing indicators at zero (the axis could be anywhere at this point). Upon rotating the part (see the second position in the figure), the top indicator shows +.002" movement, and the bottom indicator shows −.003" movement. The differential between these two values is the linear movement of the axis (.0025") from the start position. Upon rotating the part further in the same direction (see the third position in the figure), the top indicator goes back to −.001" position, and the bottom indicator goes back to +.002" position.

The differential between these two values is .0015" axial movement. The last position in the figure, upon rotating the part further, shows that both indicators agree on .0000" movement. When using this technique, the part is rotated 360° and the observer is looking for the highest differential amount found during that rotation. This amount is the direct linear movement of the axis away from the datum axis.

In this case, the highest differential amount was .0025". Concentricity, however, is a cylindrical tolerance zone, and so the observer must double this amount so that the *actual cylinder zone* is known and compared to the drawing cylinder zone allowed. This part is an eccentric .005" cylinder (two times the actual linear deviation of the axis).

Consider the part drawing in Figure 15.3. The feature control frame states that the axis of the large diameter must be concentric to the axis of the smaller diameter (datum B) within .003" cylindrical zone (RFS is understood per rule 3 of the standard). The tolerance zone for this requirement is *a .003" diameter cylindrical zone that is perfectly concentric to datum axis B,* as shown in Figure 15.4. Note that RFS is understood with concentricity requirements per rule 3. The axis of the controlled diameter must lie within this cylindrical tolerance zone.

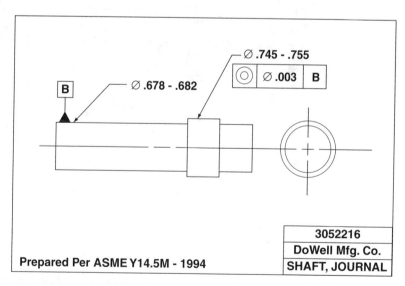

Figure 15.3 Drawing requirement that specifies a concentricity tolerance.

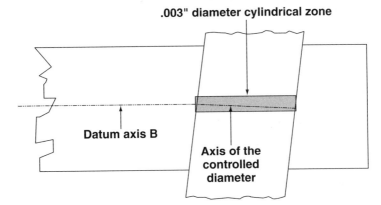

The tolerance zone is a .003" diameter cylinder that is perfectly concentric to datum axis B. The axis of the controlled diameter must lie within the cylindrical tolerance zone from end to end. Concentricity tolerance is regardless of feature size (RFS).

Figure 15.4 Tolerance zone for the part.

To inspect this part, the following equipment is required:

1. Surface plate
2. Two dial indicators that discriminate to 10% or less of the concentricity tolerance (preferably the same discrimination)
3. Two surface gages (or height gages) to hold the dial indicators
4. Vee block to contact the datum diameter
5. Two precision parallels (for elevation purposes)

STEP 1 Carefully rotate and slide the equipment onto the surface plate.

STEP 2 Mount the vee block on top of the two precision parallels (in this setup) to elevate the part. In some cases, the elevation may not be needed, but the indicators must be set at top and bottom or side to side 180° apart for the inspection.

STEP 3 Bring the two opposing indicators into a position such that they are 180° apart and each indicator is set at zero (see Figure 15.5).

STEP 4 Rotate the part 90° (or in smaller increments) and take a differential reading, as shown in Figure 15.6.

STEP 5 Rotate the part further and continue to take differential readings until the part has been rotated a full 360°, as shown in Figure 15.7. The largest differential reading is equal to one-half the actual cylindrical concentricity error. Multiply the largest differential reading by 2. If the actual cylindrical zone is equal to or smaller than the allowed cylindrical zone, the part is acceptable.

Figure 15.5 Two opposing indicators have been brought in contact with the controlled diameter and set at zero.

Figure 15.6 The part is rotated 90° and a differential reading is obtained.

Figure 15.7 Continued rotation establishes more differential readings.

CONCENTRICITY VERIFIED BY TOTAL RUNOUT

Differential measurement is used to measure the actual eccentricity of the axis. When an actual value is not needed (accept or reject is the only criteria), runout may be used to verify concentricity.

If the total runout of the controlled diameter (FIM value) is equal to or less than the cylindrical zone for concentricity, the controlled diameter is somewhere within concentricity tolerance. The observer cannot state the actual cylindrical zone, but the accept or reject decision can be made. Only total runout (refer to Chapter 12) can be used in this case. Circular runout methods will not suffice.

PRECISION-SPINDLE METHOD

Another example of the versatility of a computer-assisted precision spindle is the fact that polar graphs can be used for accurate measurements of concentricity. The part shown in Figure 15.8 has three outside diameters. There is a concentricity requirement for the part where the smallest outside diameter is the datum feature, and the next-larger diameter has a concentricity requirement. Figure 15.9 shows that the probe of the precision spindle has been set up to graph the datum diameter. Once the datum diameter

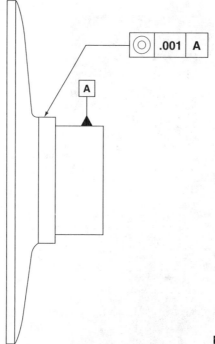

Figure 15.8 A concentricity callout.

Figure 15.9 The datum diameter is traced with the probe of the precision spindle. Courtesy of Federal Products Corp.

Figure 15.10 The controlled diameter is traced with the probe of the precision spindle. Courtesy of Federal Products Corp.

has been polar-graphed, the controlled diameter (shown in Figure 15.10) can then be graphed. The polar graph can then be used to measure the eccentricity of the axis of the controlled diameter to the axis of the datum diameter.

REVIEW QUESTIONS

1. Which of the following geometric tolerances can be used as coaxial controls: (a) runout, (b) concentricity, (c) position, (d) all of these?
2. Concentricity, when applied to an outside diameter of a shaft, automatically controls the taper of that diameter. *True* or *False?*
3. The term *verified* as used in the chapter means that the control is not necessarily directly measured, but the observer knows that it is within tolerance. Which geometric symbol can be used to verify concentricity?
4. Concentricity is easily measured using one dial indicator (or probe). *True* or *False?*
5. Which of the following terms relates closely to the measurement of concentricity?
    (a) 360° rotation is required during measurement
    (b) The axial location must be measured
    (c) Bonus tolerances are not applicable
    (d) All of these

Chapter 15 Concentricity

6. The deviations of the controlled axis from the datum axis are best measured using which technique: (a) FIM, (b) direct measurement, (c) differential measurement, (d) wall thickness variations?
7. The tolerance zone for concentricity is (a) a cylinder at MMC, (b) two parallel lines, (c) two parallel planes, (d) a cylinder, RFS.
8. Concentricity is an axis-to-axis relationship. *True* or *False?*
9. Which of the following controls is superior to concentricity for rotating parts: (a) position, (b) cylindricity, (c) runout, (d) parallelism?
10. The result of concentricity measurement on an inspection report is a FIM value. *True* or *False?*

16

Position Tolerances

OBJECTIVES

Upon completion of this chapter, the reader should be able to:

1. Understand and inspect a wide variety of position tolerance applications.
2. Apply graphical inspection analysis (paper gaging) to position tolerances.
3. Convert position tolerance diameters into coordinate ($\pm$) tolerances.
4. Convert actual position measurements into actual cylindrical zones.
5. Understand the application of functional gages for positional tolerances at maximum material condition (MMC).
6. Understand bonus tolerances and additional tolerances that can be achieved in certain applications.
7. Understand how and when position tolerances control perpendicularity.
8. Understand the proper methods for using coordinate measuring machines for measuring position tolerances.

INTRODUCTION AND APPLICATIONS

Position tolerance is the most widely used geometric tolerance symbol. Position is a location tolerance that applies to the center of size features (for example, locates the axis of a hole or shaft or the centerplane of a boss or slot). There are many position tolerance applications, including control of single size feature locations, coaxial features, multiple pattern locations, and many more. Mating parts that are fastened together with rivets, bolts, pilot pins, and other fasteners require that the features be located within a given location for proper fit. Position tolerance, per ASME Y 14.5, not only guarantees proper fit of mating parts, but also allows (due to worst-case tolerances) *bonus tolerances* that could not be used in older *coordinate ± tolerancing methods*. Guaranteed fits and bonus tolerances are one of the best combinations offered a manufacturer through applications of position tolerancing.

As indicated in Chapter 15, position tolerances can also be used to control coaxial relationships (concentricity) of nonrotating parts applied at MMC (for parts that must simply go together), least material condition (LMC) for special applications such as controlling location and wall thickness, or RFS [for direct coaxial location regardless of size (or concentricity)]. There are several different applications and measurement methods for position tolerancing, many of which will be discussed in this chapter.

KEY FACTS

1. The primary purpose of position tolerances is to *locate the center of size features* of various kinds, such as holes, pins, bosses, and tabs.
2. Position tolerances always require *datums*. In some cases, the datums are intrinsic (for example, hole to hole).
3. Position tolerance can be used for simple *locations, symmetry* applications, or *coaxiality* relationships.
4. Position tolerances for location *require basic dimensions* to define the perfect (true) position.
5. Position tolerances applied at MMC or LMC can allow the manufacturer to make use of *bonus tolerances* without loss of function of the part.
6. Position tolerances applied at MMC also allow for the option of *functional gaging* methods.
7. Position tolerances specified at regardless of feature size (RFS) cannot be functionally gaged, nor can bonus tolerances be used.
8. Position tolerances do not apply to surfaces or edges of a size feature (unless the term BOUNDARY is stated under the feature control frame). They apply only to the center of size features (for example, axes and centerplanes).

COORDINATE (±) DIMENSIONING VERSUS POSITION TOLERANCING

Some drawings use the old coordinate system of locating features. An example of this system is shown in Figure 16.1. The location of the hole must be 2.200″ ± .010″ from the bottom edge of the part and 3.250″ ± .010″ from the left edge of the part. Although the drawing appears to be simple, in comparison to other methods, it leaves a lot to be desired.

There are many problems associated with the old coordinate system described in Figure 16.1. Table 16.1 describes some of these.

As shown in Figure 16.2, the tolerance zone for the old coordinate system results in a square zone for a round hole. The square zone (±.010″ in each direction) is also

Coordinate (±) Dimensioning Versus Position Tolerancing

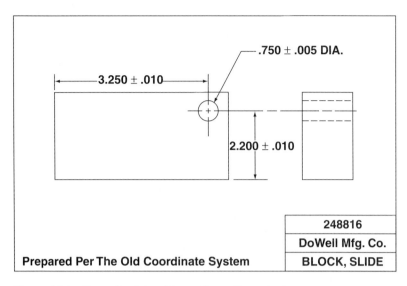

Figure 16.1 Example of the old coordinate dimensioning system.

Table 16.1 Problems with the Old Coordinate System

Subject	Problem	Effect
Datums	The edges of the part are not defined as datums.	The observer can interpret and contact these edges in any manner.
Tolerance zone shape	The coordinate system results in a square tolerance zone.	Square tolerance zones limit the amount of tolerance and are not practical for locating round holes.
Amount of tolerance	The square tolerance zone reduces usable tolerances by 57% *and* is always RFS.	Good products get rejected, and bonus tolerances cannot be used even if they are appropriate.
Two dimensional	The old system is two-dimensional and often does not control the location of the hole throughout its depth.	Products that are in location at one end, but out of perpendicularity, get accepted and do not assemble.
No interface description	If one of the faces perpendicular to the hole were an interface, it would not be described.	Production or inspection may not control the hole to the proper face.
Functional gaging	One cannot use a functional gage in RFS applications.	All parts must be measured.
Tolerance accumulation	When applied to a chain of holes, tolerances are accumulative.	Designers are often forced into tightening tolerances to avoid tolerance buildup.

234 Chapter 16 Position Tolerances

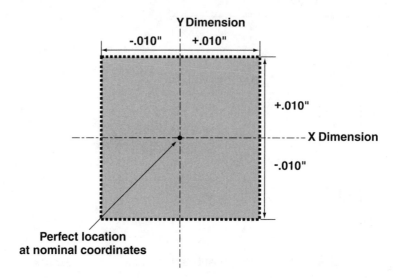

Figure 16.2 The tolerance zone in the old system is a square.

regardless of feature size in all cases. Many production and inspection problems result from the old coordinate system of locating size features. Position tolerance defines a tolerance zone for most hole locations as a cylindrical zone located by basic dimensions. These tolerances also give the designer the flexibility to define applications that must be RFS or can be MMC or LMC. In the cases of MMC and LMC applications, bonus tolerances can be achieved depending on the actual size of the feature being controlled.

A comparison of the two different tolerance zones is shown in Figure 16.3. Some interesting differences are easily seen.

1. The square zone allows .0141″ deviation from true position at the corners of the zone, but only .010″ deviation in *X* and *Y* directions. This is illogical because a round hole is being located. If .0141″ deviation is allowed in one direction, it should be allowed in all directions.
2. The round zone allows the same amount of deviation in all directions as the square zone allowed in four corners. Therefore, the virtual size of the hole, using either zone, is unchanged. Since this is true, converting to the round zone gave the producer 57% more tolerance area (even if the round zone is still RFS).
3. The new round (cylindrical) tolerance zone, depending on the application, may be designated at MMC. In this case, the actual size of the produced hole may be such that even more (bonus) tolerance for hole location can be achieved.
4. The cylindrical tolerance zone is located by basic dimensions that have no tolerance. Basic dimensions do not allow tolerance buildup over chains of holes; therefore, designers are not forced to tighten hole location dimensions unnecessarily.
5. The cylindrical tolerance zone is three dimensional, so perpendicularity must also be controlled.

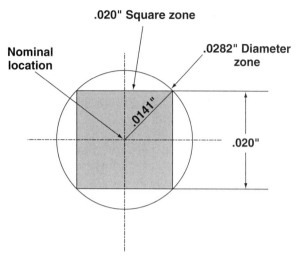

Figure 16.3 The cylindrical tolerance zone allows more tolerance than the square zone.

6. The basic dimensions begin at specified datums that are defined in terms of how they must be contacted.

To summarize, the use of geometric tolerances for positioning size features is a much more efficient system that allows more tolerances, better definition of datums, functional inspection, and many more benefits. The drawing in Figure 16.4 shows the

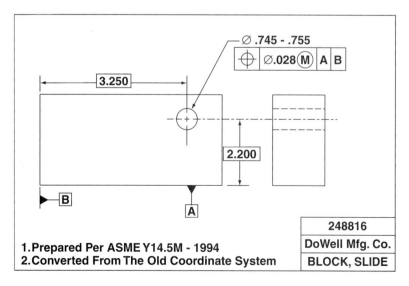

Figure 16.4 The same part dimensioned using ASME Y 14.5M, 1994 standard.

previous part redrawn using ASME Y 14.5M, 1982 standard, for position tolerancing. This method seems somewhat more complicated, but it provides a much better approach to tolerancing the part based on function. The old coordinate system, although it seems simple, can cause counterproductive results.

CONVERSION OF CYLINDRICAL ZONES TO COORDINATE ± TOLERANCES

Position tolerance zones (for control of cylindrical features) are always stated in terms of the diameter of the cylindrical zone where the axis of the feature must lie. At times, for fast inspection, the only interest of the observer is whether or not the feature lies within the zone. In these cases, the cylindrical tolerance zone can be converted into its equivalent plus or minus tolerances and measured directly on each coordinate. The equivalent plus or minus tolerance for a cylindrical tolerance zone can be found by finding one-half of the largest perfect square that will fit inside the applicable cylindrical zone. The formula for finding the equivalent plus or minus tolerances for a cylindrical zone is shown in Figure 16.5.

$$\text{Formula: } \frac{\text{Diameter of zone} \times .7071}{2}$$

$$\text{Example: } \frac{.010'' \text{ diameter} \times .7071}{2}$$

The tolerance for each coordinate is ± .00353".

Figure 16.5 The formula for finding the equivalent plus or minus tolerance for a given cylindrical tolerance zone. (Plus and minus tolerances found apply equally for both coordinates.)

For example, the equivalent plus or minus tolerance for a .010" cylindrical zone would be plus or minus .00353" because the largest square that will fit inside a .010" cylindrical zone is a .00707" square. The square then divided by 2 allows the axis of the feature to vary plus or minus .00353" from true position. Another example is that a .030" cylindrical zone converts to a coordinate plus or minus tolerance of .0106".

CONVERSION OF COORDINATES TO CYLINDRICAL ZONES

In many cases, the observer must evaluate position tolerances and report the actual cylindrical zone compared to the specified zone. Upon inspecting the actual location of each basic coordinate, the observer should record the deviation of each measured coordinate from the basic dimension for that coordinate. Then, using the formula given in Figure 16.6, the observer can compute the *actual cylindrical zone* for the axis of the feature.

$$\text{Formula: Actual diameter} = 2\left(\sqrt{\Delta X^2 + \Delta Y^2}\right)$$

$$\text{Example: Actual diameter} = 2\left(\sqrt{.003''^2 + .004''^2}\right)$$

$$= .010'' \text{ diameter}$$

Figure 16.6 Formula for computing the actual cylindrical zone from inspection measurements.

For example, if the actual deviations from basic location are .002″ in the X direction and .005″ in the Y direction (ignore plus or minus signs), the actual cylindrical zone for the axis is .0108″ diameter. Another method for finding the actual cylindrical zone is to use the table in Figure 16.7. This table has been prepared based on the formula in Figure 16.6. In most cases, the observer must report the actual cylindrical zone compared to the required cylindrical zone for position tolerances.

BONUS TOLERANCES

Another important aspect about inspecting position tolerances is the understanding and use of bonus tolerances when they apply. Bonus tolerances apply only when the modifiers for MMC or LMC are shown in the feature control frame. The foundation for allowing bonus tolerances in mating parts is the *virtual condition* (or worst case).

An example of virtual condition is a hole that is larger than its MMC size, and therefore has more clearance for the bolt to passthrough. In this case the hole can be farther out of location due to this clearance and still allow the bolt to go through. The virtual condition is the worst case, depending on the location and the size of the hole.

The bonus tolerance is the additional amount of position tolerance allowed because the hole is larger than MMC size (or a shaft is smaller than MMC size). The bonus, then, is the difference between the MMC size and the actual size of the feature.

An example MMC application is shown in Figure 16.8. The feature control frame states that the hole must be in position within a .010″ cylindrical tolerance zone (when the hole is produced at MMC size) with respect to datums A (primary), B (secondary), and C (tertiary). If the hole is produced at MMC size, no bonus position tolerance is allowed. If the hole is produced larger than its MMC size, that amount can be added to the stated position tolerance as bonus tolerance. The table shown in Figure 16.8 shows examples of bonus tolerances allowed. The total position tolerance is the stated position tolerance plus the allowed bonus tolerance. Notice, in the table, that no matter what bonus tolerance has been applied to the part (based on the actual size of the produced hole), the virtual size remains the same. In position tolerance applications that specify MMC, the virtual condition of the feature is the primary consideration.

Y Deviation											
.020	.0400	.0402	.0404	.0408	.0418	.0418	.0424	.0431	.0439	.0447	.0456
.019	.0380	.0382	.0385	.0388	.0393	.0398	.0405	.0412	.0420	.0429	.0439
.018	.0360	.0362	.0365	.0369	.0374	.0379	.0386	.0394	.0402	.0412	.0422
.017	.0340	.0342	.0345	.0349	.0354	.0360	.0368	.0376	.0385	.0394	.0405
.016	.0321	.0322	.0325	.0330	.0335	.0342	.0349	.0358	.0367	.0377	.0388
.015	.0301	.0303	.0306	.0310	.0316	.0323	.0331	.0340	.0350	.0360	.0372
.014	.0281	.0283	.0286	.0291	.0297	.0305	.0313	.0322	.0333	.0344	.0356
.013	.0261	.0263	.0267	.0272	.0278	.0286	.0295	.0305	.0316	.0328	.0340
.012	.0241	.0243	.0247	.0253	.0260	.0268	.0278	.0288	.0300	.0312	.0325
.011	.0221	.0224	.0228	.0234	.0242	.0250	.0261	.0272	.0284	.0297	.0311
.010	.0201	.0204	.0209	.0215	.0224	.0233	.0244	.0256	.0269	.0283	.0297
.009	.0181	.0184	.0190	.0197	.0206	.0216	.0228	.0241	.0254	.0269	.0284
.008	.0161	.0165	.0171	.0179	.0189	.0200	.0213	.0226	.0241	.0256	.0272
.007	.0141	.0146	.0152	.0161	.0172	.0184	.0198	.0213	.0228	.0244	.0261
.006	.0122	.0126	.0134	.0144	.0156	.0170	.0184	.0200	.0216	.0233	.0250
.005	.0102	.0108	.0117	.0128	.0141	.0156	.0172	.0189	.0206	.0224	.0242
.004	.0082	.0089	.0100	.0113	.0128	.0144	.0161	.0179	.0197	.0215	.0234
.003	.0063	.0072	.0085	.0100	.0117	.0134	.0152	.0171	.0190	.0209	.0228
.002	.0045	.0056	.0072	.0089	.0108	.0126	.0146	.0165	.0184	.0204	.0224
.001	.0028	.0045	.0063	.0082	.0102	.0122	.0141	.0161	.0181	.0201	.0221
	.001	.002	.003	.004	.005	.006	.007	.008	.009	.010	.011

X Deviation

Figure 16.7 Table for converting the actual deviations from coordinate measurements to the actual cylindrical zone.

SINGLE FEATURE POSITION TOLERANCE

When applying position tolerance to single feature locations, such as the position tolerance for the holes in Figure 16.9, the application is fairly straightforward. The feature control frame states that the axis of each hole must lie within a .010″ cylindrical tolerance zone (regardless of feature size) with respect to primary datum *A*, secondary datum *C*, and tertiary datum *B*. It is up to the observer to inspect the coordinate locations of the axis of each hole and then determine whether or not the axis of the hole lies within the stated cylindrical tolerance zone (that has been located by basic dimensions).

Single Feature Position Tolerance

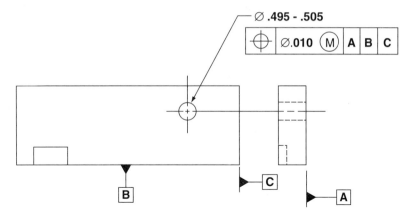

Actual hole Size (")	Bonus tolerance Allowed (")	Total Position Tolerance (")	Virtual Size (")
.495 (MMC)	0	.010	.485
.499	.004	.014	.485
.502	.007	.017	.485
.505 (LMC)	.010	.020	.485

Figure 16.8 Example of bonus tolerances for a position tolerance applied at MMC.

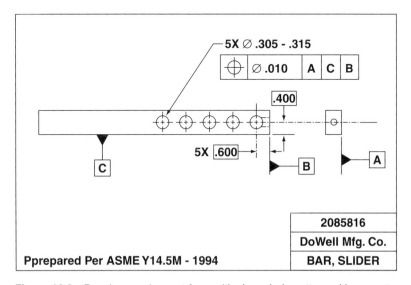

Figure 16.9 Drawing requirement for positioning a hole pattern with respect to three datums.

Chapter 16 Position Tolerances

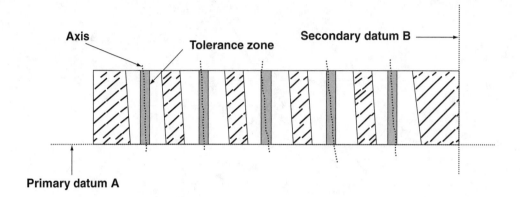

The tolerance (for each hole) is a .010" diameter cylinder regardless of the size of the hole. The .600" basic dimensions locate the tolerance zones from secondary datum plane B. Each tolerance zone is perpendicular to primary datum plane A, and each zone is .400" basic from tertiary datum plane C.

Since the tolerance zones are located by basic dimensions, there is no build up of tolerances for the five "chain-dimensioned" holes. The axis of each hole must lie within its tolerance zone from end to end.

Figure 16.10   Tolerance zones for the part.

The tolerance zone for each hole is *a .010" diameter cylinder, regardless of the hole size, that is located by basic dimensions and is perfectly perpendicular to datum plane A and perfectly parallel to datum plane B*, as shown in Figure 16.10. The acceptance of the part must be with respect to the tolerance zone, not the basic dimensions.

The inspection setup for this part must contact all three datums properly and then evaluate the location of the axis of the hole (finding the actual cylindrical zone where the hole lies). If the datums were only the outside edges of the part, the location of the hole (within the cylindrical tolerance zone) would be the only requirement. Since the primary datum is a face that is perpendicular to the hole, the perpendicularity of the hole to the primary datum must also be inspected.

The following equipment is necessary for inspection in an open setup on a surface plate:

1. Surface plate
2. Dial indicator that discriminates to 10% or less of the total position tolerance
3. Height gage to hold the dial indicator and measure linear distances from datum planes
4. Set of gage blocks (optional for using the transfer method)
5. Largest true gage pin that can be inserted into each controlled hole
6. Two precision parallels (right-angle plates or square blocks) to contact the datums

The example shown in Figure 16.9 has several holes. The following method focuses on the end hole nearest datum *B* to provide an example. All the holes could be inspected for the same coordinates using the same setup.

STEP 1 Carefully rotate and slide the inspection equipment onto the surface plate.

STEP 2 Insert the gage pin selected into the controlled hole in a manner such that the pin will not protrude past datum *A*.

STEP 3 Mount the primary, secondary, and tertiary datums onto the accessories in a manner such that the gage pin is horizontal to the surface plate, as shown in Figure 16.11.

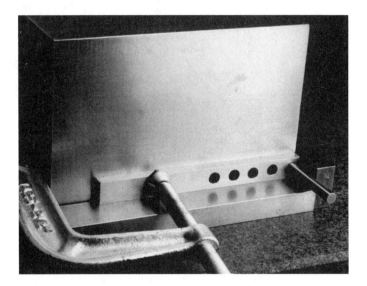

Figure 16.11 The part is mounted on the datums and ready for inspection of the first hole.

STEP 4 Measure the distance from datum *C* (the parallel) to the top of the gage pin and subtract one-half the pin size, as shown in Figure 16.12. Make sure that this measurement is made next to the hole. Record the actual measurement for later use. An option here is to stack gage block stacks that are the basic dimension plus one-half the gage pin size used. In this manner, the height that should result at the top of the gage pin can be transferred using a surface gage instead of a height gage (this avoids having to read the vernier scale on the height gage).

STEP 5 Rotate the setup 90° so that the part is perpendicular to the surface plate (datum *B* is nearest to the surface plate). Measure the coordinate location from datum *B* (the surface plate) to the top of the gage pin and subtract one-half the gage pin size (see Figure 16.13). Record the measurement for later use.

242 Chapter 16 Position Tolerances

Figure 16.12 The first coordinate location is measured on the part using a gage pin.

Figure 16.13 The setup is rotated 90°, and the other coordinate dimension is measured.

STEP 6 Evaluate the coordinate measurements made to find the position cylinder of the produced hole at that end. First, take the two linear dimensions recorded and find the differences between them and the basic dimension for that coordinate. Once again, the observer is looking only for the deviations from true position (defined by the basic dimensions). Next, compute the actual cylindrical zone for these two coordinates using the formula or table specified earlier in this chapter. If the actual cylindrical zone is larger than the specified zone, the part is not acceptable (at that end of the hole).

STEP 7 For the perpendicularity inspection, the observer has two options. One is to repeat the previous steps taking measurements at the other end of the hole (next to the primary datum) and evaluating these measurements with respect to the same basic dimensions and tolerance zone. The other option is to directly measure the perpendicularity of the hole with respect to datum plane *A* and then relate the lean of the hole to the location of the axis that has already been measured. In either case, perpendicularity error from datum *A cannot allow the axis of the hole to be outside of the position tolerance zone at either end.*

COAXIAL FEATURES

Position tolerances are often used (instead of concentricity or runout) to control the coaxiality of two cylindrical features (outside diameters or inside diameters, or both). In these cases, the function of the part is not rotation. If the part rotated in assembly, concentricity or runout would have been applied. Figure 16.14 shows an example part drawing in which position has been used for coaxial control. The feature control frame

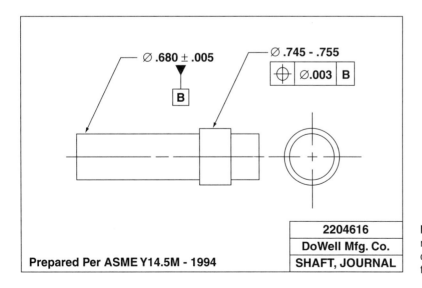

Figure 16.14 Drawing requirement for coaxial control using position tolerancing.

244   Chapter 16   Position Tolerances

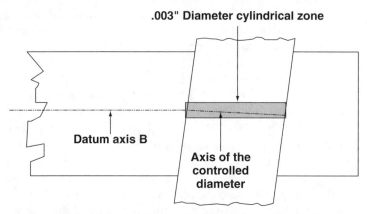

Figure 16.15   Tolerance zone for the part.

states that the axis of the controlled diameter must be coaxial to the axis of datum *B* (regardless of feature size) within a .003″ cylindrical zone (regardless of feature size). The tolerance zone for this requirement is *a .003″ diameter cylinder that is perfectly concentric to datum axis B*, as shown in Figure 16.15. The axis of the controlled diameter must lie within the cylindrical tolerance zone.

There is not a lot of difference between position coaxial at RFS and concentricity. The drawing shown in Figure 16.14 is the same part used in Chapter 15 (concentricity) for which differential measurements were made to evaluate concentricity. The position tolerance at RFS is measured the same way, except that constant rotation during the measurement is not a requirement. Refer to the inspection method for concentricity on this part in Chapter 15.

POSITION: COAXIAL AT MMC

If the previous part were called out at MMC instead of RFS, as shown in Figure 16.16, a functional gage could be designed and used to inspect the part. The feature control frame states that the axis of the controlled diameter (at MMC) must be coaxial to the axis of the datum diameter (at MMC) within .003″ cylindrical zone. The tolerance zone, in this case, is a *.003″ diameter cylinder (when the controlled feature is produced at MMC size) that is perfectly concentric to datum axis B. As the controlled diameter produced departs from MMC size, a bonus tolerance allows the cylindrical zone to grow. As the datum diameter departs from MMC size, an additional tolerance allows the tolerance zone to move away from datum axis B*, as shown in Figure 16.17. More tolerance is achievable if either the controlled feature or the datum diameter (or both) departs from MMC size.

Position: Coaxial at MMC   **245**

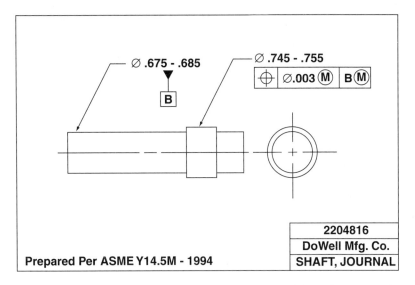

Figure 16.16   The drawing has been changed to a coaxial position tolerance specified at MMC.

The tolerance zone is a .003" diameter cylinder that is concentric to datum axis B. The axis of the controlled diameter must lie within the tolerance zone from end to end.

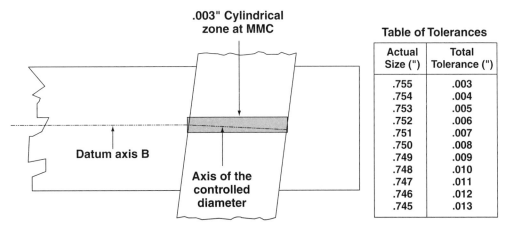

Actual Size (")	Total Tolerance (")
.755	.003
.754	.004
.753	.005
.752	.006
.751	.007
.750	.008
.749	.009
.748	.010
.747	.011
.746	.012
.745	.013

If the controlled feature departs from MMC size, a bonus tolerance is allowed that is equal to the difference between the MMC size and the actual size. As the datum feature departs from MMC size, additional tolerance is allowed, for example the tolerance zone can move or shift from the datum axis.

Figure 16.17   Tolerance zone for the part.

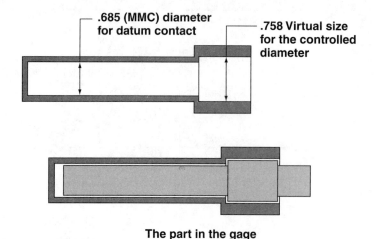

Figure 16.18 Functional gage design for the coaxial position tolerance at MMC.

An example of a functional gage for this part (with the part inside) is shown in Figure 16.18. The diameter in the gage that contacts datum *B* has been made at the MMC size of datum *B* within gage tolerances. The diameter of the gage that contacts the controlled diameter is at the virtual condition of the controlled diameter within gage tolerances. Both of the functional gage diameters are coaxial to each other within gage tolerances. Gaging tolerances are typically 10% of the part tolerance or less.

When using functional gages, the observer must understand that, due to gagemaker tolerances and wear allowances, the gage may reject a borderline good part at times. In these cases, it is up to the observer to measure the position tolerance in open setup before making the acceptance decision. It is also understood that all a functional gage checks is the virtual size of the part (worst case). The size limits of the diameters must be inspected separately.

Another example of coaxial control is the part drawing shown in Figure 16.19. In this case, the feature control frame states that the three holes must be in line (coaxial) to one another within .005″ cylindrical tolerance zone (when the holes are produced at MMC size). The tolerance zones for this requirement is *a .005″ diameter cylinder (when the hole is at MMC size) for each hole where the axis of the hole must lie. The .005″ diameter cylinder zones are perfectly concentric to each other*, as shown in Figure 16.20.

If any hole is produced larger than MMC size, the tolerance zone for that hole is larger (does not move), allowing more area for acceptance of the part. A functional gage could also be used on this part. The functional gage, in this case, is simply a gage pin that is long enough to go through all three holes and is at the virtual size of the holes. Refer to Figure 16.21 for an example of the functional gage.

Feature Pattern Locations   **247**

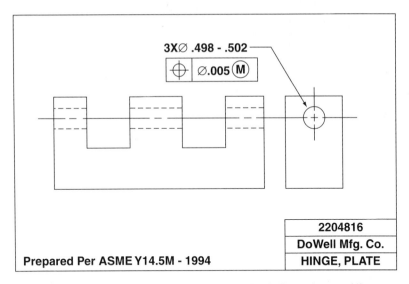

Figure 16.19 Drawing requirement for three holes in line to be coaxial (positioned in line) when they are at MMC.

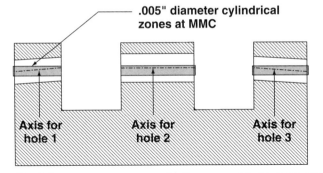

Actual Hole Size (")	Bonus Tolerance (")	Total Tolerance (")
.498	.000	.005
.499	.001	.006
.500	.002	.007
.501	.003	.008
.502	.004	.009

The tolerance zones are perfectly concentric to each other and they are .005" diameter cylinders when the hole is at MMC size. If any hole is larger than its MMC size, the zone for that hole increases by an amount equal to the difference between the actual hole size and the MMC size.

Figure 16.20 Tolerance zone for the part.

FEATURE PATTERN LOCATIONS

In many design applications, there are patterns of holes, pins, or the like that must be in position in order to fit the mating part. Patterns may be shown in a variety of methods for position on the drawing, beginning with the following example. Whenever a pattern of features (such as a hole pattern) has a position tolerance applied that does not

248   Chapter 16   Position Tolerances

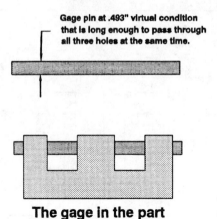

Figure 16.21 Functional gage pin for the three-hole coaxial requirement at MMC.

specify datums, the datum is *intrinsic (or the features to each other)*. The drawing in Figure 16.22 shows an example of a hole pattern with an intrinsic datum. The feature control frame states that the holes must be in position tolerance to one another within .010″ diameter cylindrical tolerance zones (regardless of feature size).

The tolerance zones for this requirement consist of *four .010″ diameter cylindrical zones that are 2.200″ (basic) apart in both directions*, as shown in Figure 16.23. These zones are .010″ diameter regardless of the size of the holes. The axis of each hole must lie within its cylindrical tolerance zone. It is also important to note that these zones con-

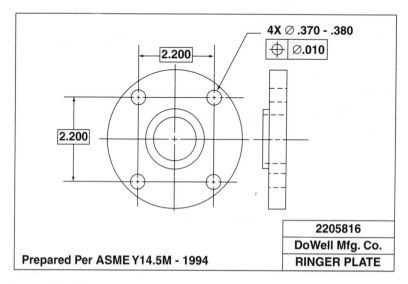

Figure 16.22 Drawing requirement for locating a pattern of holes to one another (intrinsic datum).

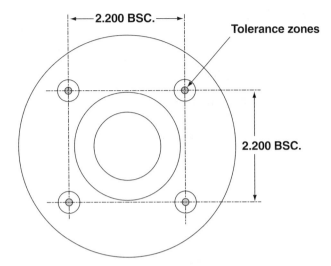

The tolerance zones are four .010" diameter cylinders that are 2.200" basic apart. These zones are regardless of feature size; therefore, no bonus tolerance is allowed. The axis of each hole must be within its tolerance zone.

Figure 16.23   Tolerance zone for the part.

trol only the position of the axis of the holes *to one another*, not the location of the pattern of holes on the part.

Feature patterns are more complex than the one-hole examples covered earlier (especially the example shown in Figure 16.22). This is so because the real requirement here is that the holes must be related during inspection to one another, and in the final analysis the only important thing is the *best-fit relationship among them*.

Each hole has its own cylindrical tolerance zone that is located by the basic dimensions from hole to hole. The other problem is that there is no set starting point for the measurement (such as outside-edge datums). The primary consideration is that all the holes be in their respective tolerance zone (anywhere) at the same time.

The other problem is that there are only two ways to properly evaluate this part with accurate results: (1) functional gaging and (2) graphical inspection analysis (GIA) (often called *paper gaging*). Coordinate measurements (*X* and *Y*) will often fool the observer into thinking the pattern is out of tolerance when it really isn't. It is only the simultaneous measurement of the holes in their zones (best fit) that can make the acceptance decision.

Only functional gages or graphical inspection analysis can help the observer make this decision. The other problem is that, unless MMC is specified, functional gages cannot be designed and used on the part. To cover the inspection of this part, the following technique will not only inspect the part, but also introduce GIA methods for evaluating the results.

SURFACE-PLATE SETUP

The following equipment is required for open-setup measurement of the part on the surface plate.

1. Surface plate
2. Dial indicator that discriminates to 10% or less of the total position tolerance
3. Height gage to hold the dial indicator and make linear measurements
4. The largest gage pins that can be inserted into each of the controlled holes
5. Precision parallel
6. Right-angle plate (or precision-square block)
7. Simple compass
8. Pad of grid paper
9. Sheet of clear Mylar (or transparency paper) to prepare an overlay gage

STEP 1 Carefully rotate and slide the equipment onto the surface plate.

STEP 2 Insert the gage pins selected into all four holes. Number the holes on a piece of sketch paper 1, 2, 3, and 4, associating hole 1 with some other feature on the part (or marking it in a manner that will not damage the part).

STEP 3 Mount the part onto a precision parallel in a manner such that two of the gage pins rest on the parallel, the back of the part rests against the right-angle plate, and hole 1 is in the upper-right corner, as shown in Figure 16.24.

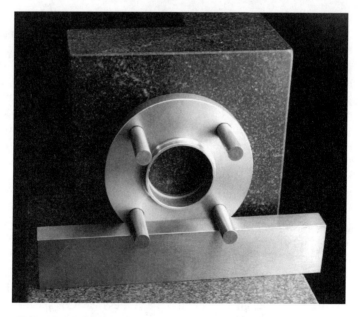

Figure 16.24 The part is mounted on two of the gage pins and supported in the back by a right-angle plate.

STEP 4 Measure the distance from the top of the parallel (where the pins are resting) (see Figure 16.25a) to the top of the opposite pin, as shown in Figure 16.25b. Do this for both sets of holes (hole 3 to 4, then hole 2 to 1). From each of these measurements, subtract one-half the size of the pins at each end and record the value. This is the distance between the two hole axes in that direction.

STEP 5 Rotate the part 90° and repeat step 4 to measure the distance over both sets of pins (hole 2 to 3 and hole 1 to 4) in the other direction. Record the axis-to-axis distance as found by subtracting one-half the size of the pins at each end from the overall measurement.

STEP 6 On a piece of paper, record the distances between all the holes, keeping them numbered so that the holes can be found (or reinspected) later if necessary.

STEP 7 Prepare an inspection report similar to the one in Figure 16.26 so that all the information can be used to evaluate the part to its position tolerance.

The blocks on the report for X coordinate and Y coordinate difference can be completed at this point by splitting the measured dimension between the two holes. For this example, let's say that the following measurements were made:

Hole 3 to 4 (after subtracting one-half the size of the pin) was 2.196.

Hole 2 to 1 (after subtracting one-half the size of the pin) was 2.196.

Hole 2 to 3 (after subtracting one-half the size of the pin) was 2.207.

Hole 1 to 4 (after subtracting one-half the size of the pin) was 2.196.

For the purpose of the inspection report, the measurement can be evenly split between the two holes in coordinate locations (for example, if the overall measurement is .002″ longer than basic, each hole can be plotted as .001″ beyond basic). The GIA technique will check for best fit regardless of how the error is distributed. This is one example of how GIA compensates for measurement coordinate differences automatically. Without GIA, the part could be falsely rejected based on simple coordinates.

STEP 8 Prepare a graphical inspection analysis (paper gage) to evaluate the part using the following steps:

(a) On a piece of grid paper, in four square locations, plot crosses that represent the perfect (basic dimension) positions of the four holes. Label the four holes 1, 2, 3, and 4, beginning with 1 for the upper-right corner hole. Then plot a dot for each hole that represents the actual X and Y deviation from true position (using the inspection report). This plot is called a *data graph* (refer to Figure 16.27). (*Note*: In this example, each square on the data graph equals .001″.)

(b) Take a sheet of Mylar (or clear transparency paper) and lay it on top of the data graph. With a compass, insert the needle point of the compass into

252 Chapter 16 Position Tolerances

(a)

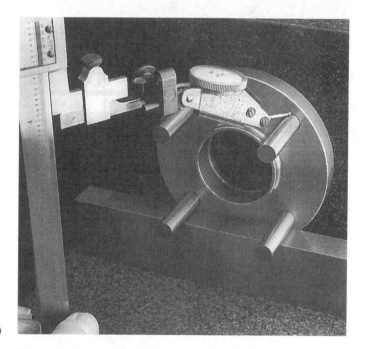

(b)

Figure 16.25 (a) Zeroing the indicator on top of the parallel, (b) Zeroing the indicator on top of the pin.

Inspection Report

Hole	Hole Size	Departure from MMC	Bonus Tolerance	X Coordinate Difference	Y Coordinate Difference	Total Position Tolerance	Actual Position Diameter per table
1	.374	.004	0*	−.002	−.001	.010*	.0045
2	.376	.006	0*	+.005	+.003	.010*	.0117
3	.372	.002	0*	−.002	+.003	.010*	.0072
4	.375	.005	0*	+.002	−.001	.010*	.0045

* Bonus tolerance is not applicable, RFS applies.

Note: Hole #2 appears to be out of position tolerance. Check with GIA to be sure.

Figure 16.26 Inspection report that shows the results of inspection.

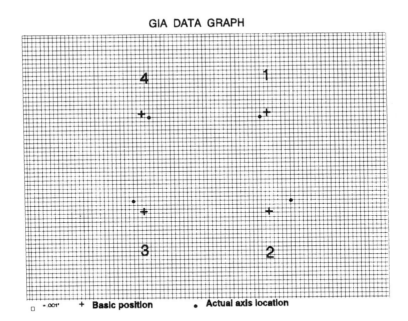

Figure 16.27 A GIA data graph has been prepared for evaluating the results.

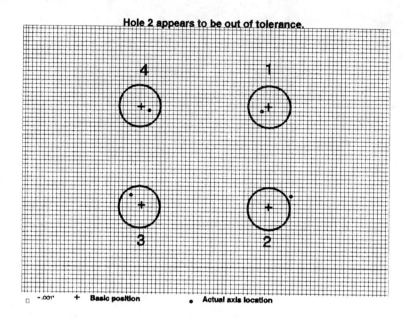

Figure 16.28 The overlay gage has been prepared while it was transposed over the data graph.

the cross for hole 1 and draw a circle that is five lines (.005″) radius. This circle represents the .010″ diameter zone allowed for position. Holding the Mylar steady to the data graph, repeat this step for the rest of the four holes, as shown in Figure 16.28. The Mylar is called the *overlay gage*.

(c) Evaluate the paper gage results for position of the four holes. In Figure 16.28, the overlay gage has been aligned with the position crosses, and hole 2 appears to be out of its position tolerance zone. In this case, since no edge datums fix the tolerance zones in space, the overlay gage can be moved or shifted in any direction to attempt to get the four plotted axes into the four cylindrical zones at the same time. If the overlay can be shifted to get all four plots in all four zones (best fit), the part is acceptable.

(d) Shift the overlay gage for best fit in the tolerance zones. As shown in Figure 16.29, the part is acceptable upon shifting the overlay gage. [*Note:* In this case, the observer was allowed to shift the overlay gage only because the requirement for position of this pattern of holes was an *intrinsic datum (holes to each other)*.] If the requirement calls for specific datums (such as the outside edges of the part), the overlay gage must be aligned with those datums and the crosses representing true position, and the part must fall within the zones that are fixed in space.

Surface-Plate Setup 255

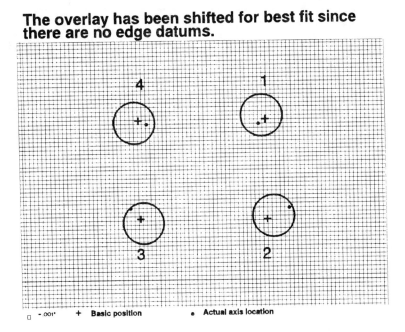

Figure 16.29 The overlay gage has been shifted for best fit and the part is acceptable.

Simple Caliper Measurement

This part could have also been inspected by using a vernier (or dial) caliper to measure the distance over the gage pins as long as the caliper was manipulated to measure the direct over-the-pin distance (see Figure 16.30). After caliper measurement, however, the observer still needs to go through the GIA procedure to evaluate the part.

Functional Gaging at MMC

If the previous part drawing specified MMC on the position tolerance for the pattern of holes, as shown in Figure 16.31, the tolerance zones would be different. The feature control frame states that the holes must be in position tolerance to one another within .010″ diameter cylindrical tolerance zones (only when the holes are at MMC size).

The tolerance zones for this requirement consist of *four .010″ diameter cylindrical zones (only when the hole is at MMC size) that are 1.400″ (basic) apart in both directions,* as shown in Figure 16.32. If the holes are produced larger than MMC size, a bonus tolerance is allowed that increases the diameter of the tolerance zone. A functional gage could be designed and used similar to the one shown in Figure 16.33.

256 Chapter 16 Position Tolerances

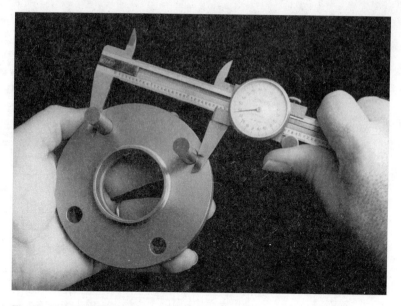

Figure 16.30 Calipers could also have been used to inspect the part, but GIA is still needed to properly evaluate the results for best fit.

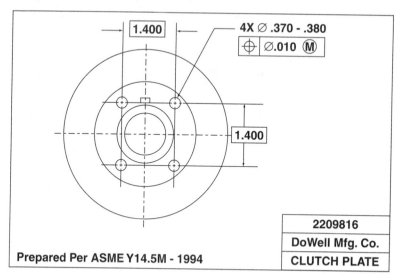

Figure 16.31 Drawing requirement has been changed to position of the pattern at MMC.

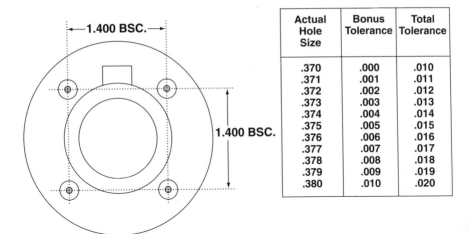

Actual Hole Size	Bonus Tolerance	Total Tolerance
.370	.000	.010
.371	.001	.011
.372	.002	.012
.373	.003	.013
.374	.004	.014
.375	.005	.015
.376	.006	.016
.377	.007	.017
.378	.008	.018
.379	.009	.019
.380	.010	.020

The tolerance zones are four cylinders that are .010" diameter only when the hole is at MMC size. If any hole is larger than MMC size, a bonus position tolerance is allowed for that hole. The axis of each hole must be within its cylindrical tolerance zone from end to end.

Figure 16.32 Tolerance zone for the part.

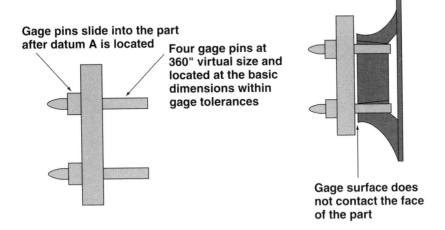

Figure 16.33 Functional gage design for the pattern of holes to each other at MMC. Notice that the gage only checks the holes to one another.

FEATURE-TO-FEATURE AND PERPENDICULARITY CONTROL

In the previous example, a pattern of holes were positioned *to one another* (intrinsic datum). A more stringent level of control is when the drawing calls for a primary datum surface on a pattern of holes (and the primary datum surface is an interface in assembly). The datum surface is perpendicular to the holes; therefore, perpendicularity of the holes to the primary datum is an added control beyond the holes to each other.

The same tolerance zones, in this case, control not only the location of the holes in the pattern to each other, but also the perpendicularity of each hole to the primary datum. The primary datum must also be contacted during the inspection. In this case, simple vernier calipers cannot be used to check the part. A layout of some sort is required. An example drawing for parts with this type of position tolerance requirement is shown in Figure 16.34.

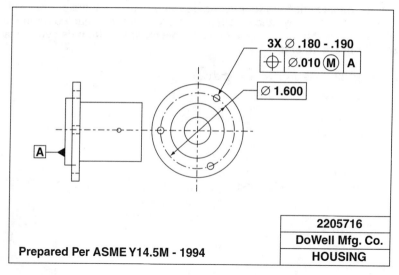

Figure 16.34 Drawing requirement for the position of a pattern of holes to each other *and* perpendicularity to an interface datum plane at MMC.

The feature control frame states that the holes must be in position within cylindrical tolerance zones of .010″ diameter (when the holes are at MMC size) with respect to datum plane *A*. (*Note:* Including datum plane *A* in this requirement also requires that the holes be perpendicular to datum plane *A* within the same .010″ cylindrical tolerance zones.) The tolerance zones for this requirement consist of *three .010″-diameter cylindrical zones (when the hole is produced at MMC size) that are equally spaced on the basic bolt circle diameter, and perfectly perpendicular to datum plane A*, as shown in Figure 16.35. The axis of each hole must lie within its tolerance zone. The tolerance zones can increase in diameter as the hole size increases from MMC size.

Feature-to-Feature and Perpendicularity Control 259

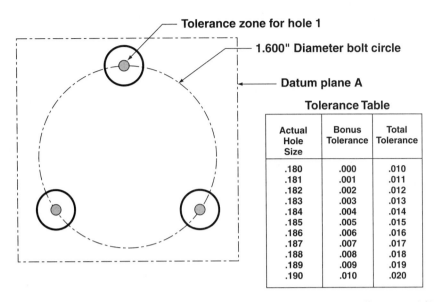

The tolerance zones are three .010" diameter cylinders that are equally spaced (120°) apart on the 1.600" basic bolt circle. The .010" diameter zone for a given hole can increase if the hole size departs from MMC size. The tolerance zones are also perpendicular to datum plane A. The axis of each hole must be within its zone.

Figure 16.35 Tolerance zone for the part.

The part in Figure 16.34 has a position tolerance at MMC on the hole pattern and perpendicularity, and yet the following methods can be used to inspect the part, first assuming that a functional gage is not affordable or appropriate for the task. After open-setup inspection and graphical inspection analysis, a functional gage design will be reviewed. To inspect the part using a surface-plate setup, the following equipment is required:

1. Surface plate
2. Dial indicator that discriminates to 10% or less of the position tolerance
3. Height gage to hold the dial indicator and make linear measurements
4. Right-angle plate (or precision-square block) to locate the primary datum
5. Precision parallels (or two equal gage block stacks) to prealign the gage pins for measurement
6. The largest gage pins that can be inserted into the holes
7. Paper gaging materials, such as grid paper, Mylar, and a compass.

STEP 1 Carefully rotate and slide the inspection equipment onto the surface plate.

STEP 2 Since this part has a basic bolt circle (instead of linear basic dimensions locating the holes to one another), linear dimensions will have to be computed for X and Y measurement. Another tool that could be used here (if the part has a datum axis) is the bolt circle chart provided in Appendix C. The bolt

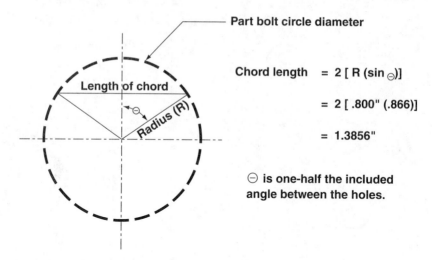

Figure 16.36 Solving for the basic chord length between the holes for measurement purposes.

circle chart, for most bolt circles, can be used directly to compute the linear *basic* dimensions from hole to hole on this and other patterns. Using the bolt circle chart and the 1.600″ basic dimension given on the drawing, the linear basic dimensions can be calculated.

Since the pattern is not related to a datum axis, the bolt circle chart is ruled out and another method should be used. In this case, the *chord dimension* between the holes has been selected. The chord length offers a direct measurement for evaluating this part. A chord length is any straight-line distance from one point on a circle to another that does not go through the center.

To calculate the chord lengths between these holes, Figure 16.36 shows the equation and the solutions. It is understood that the included angle for the holes is 120°; therefore, angle theta (θ) is equal to 60° in this problem.

STEP 3 Insert the three selected gage pins into the controlled holes so that they go through, but do not extend past datum surface A. Number the holes in a manner that will not damage the part.

STEP 4 Mount the part datum surface A on the right-angle plate (or a precision-square block) and clamp lightly. Then, choosing any two pins, use the two gage block stacks (or precision parallels) to align them vertically, as shown in Figure 16.37. Clamp the part securely.

STEP 5 Rotate the square block (or right-angle plate) 90° so that the two prealigned pins are vertical. Measure the distance from the top of one pin to the top of the other pin and add or subtract (as appropriate) the sizes of the pins to obtain the center-to-center distance between the axes of the two holes. Record the length of the chord as the distance between holes 1 and 2 (for example), as shown in Figures 16.38 and 16.39.

Feature-to-Feature and Perpendicularity Control   **261**

Figure 16.37 The part is mounted on datum *A*, aligned, and ready for the setup to be rotated 90° for inspection.

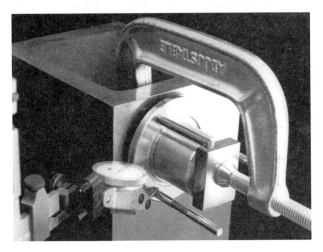

Figure 16.38 Indicating zero at top dead center of the bottom pin.

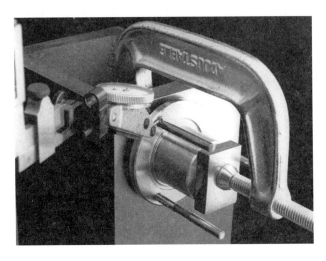

Figure 16.39 Indicating zero at top dead center of the top pin.

Chapter 16 Position Tolerances

STEP 6 While the part is still mounted on the right-angle plate, measure the perpendicularity of each hole, making sure to note the amount of lean and the exact direction of the lean of the hole to datum *A*. This can be accomplished by using a distance on the pin that is equal to the length of the axis of the hole (or, in other words, the thickness of the part).

STEP 7 Repeat step 5 after rotating the part until all three chord measurements have been made and recorded. The results should be an inspection report that has the following information:

Chord distance between holes 1 and 2

Chord distance between holes 2 and 3

Chord distance between holes 3 and 1

STEP 8 Complete the inspection report shown in Figure 16.40 so that a paper gage can be constructed to evaluate the position tolerance.

Hole No.	Hole Size (")	Departure from MMC (")	Bonus Tolerance (")	X Coordinate Difference (")	Y Coordinate Difference (")	Total Position Tolerance Allowed	Perpendicularity Error (")
1	.188	.008	.008	−.003	+.003	.018	.0036
2	.187	.007	.007	+.004	−.005	.017	.0038
3	.186	.006	.006	−.002	−.002	.016	.0032

Figure 16.40 Inspection report for the results.

STEP 9 Prepare a paper gage for the relationship, as shown in Figures 16.41, 16.42, and 16.43. Based on the finished paper gage, decide whether the part is within tolerance.

As shown in Figure 16.42, the part (specifically hole 2) is not in a cylindrical tolerance zone of .010″ diameter. If the requirement were position RFS, the part would be rejected at this point. However, the requirement is position at MMC, and so the observer should review the inspection report for the sizes of each hole to determine the amount of bonus position tolerance allowed.

In this case, the size of hole 2 (per the inspection report) is such that .007″ bonus position tolerance is allowed. The paper gage is modified in Figure 16.43 to show the allowable tolerance zones for each hole *and* the total tolerance allowed due to added bonus tolerance. Per the paper gage in Figure 16.43, the observer would accept the part because it is within position tolerance.

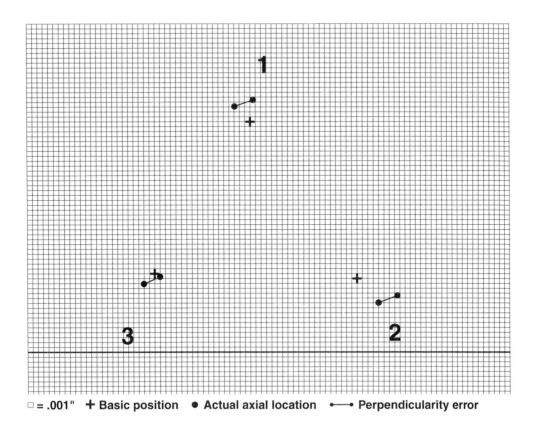

Figure 16.41 A data graph is prepared to show the results (perpendicularity error is also plotted).

PATTERN LOCATION, FEATURE TO FEATURE, AND PERPENDICULARITY

This example of position tolerancing is again more complex than the previous examples and more difficult to produce and inspect. In Figure 16.44, the part has a position tolerance that controls four different aspects about the two through holes: first, the location of the holes with respect to primary datum A, secondary multiple datums B and C, and tertiary datum D, using all the basic dimensions.

Second, the hole-to-hole location is controlled. Third is the perpendicularity of both sets of through holes to datum A. The last control is the coaxiality of each set of through holes. All these controls are built into the .005" cylindrical tolerance zones located by the basic dimensions. The two cylindrical tolerance zones are perfectly perpendicular to datum A, perfectly parallel and located to datums B and C, and perfectly parallel and located to datum D. The zones also extend through all three holes.

264 Chapter 16 Position Tolerances

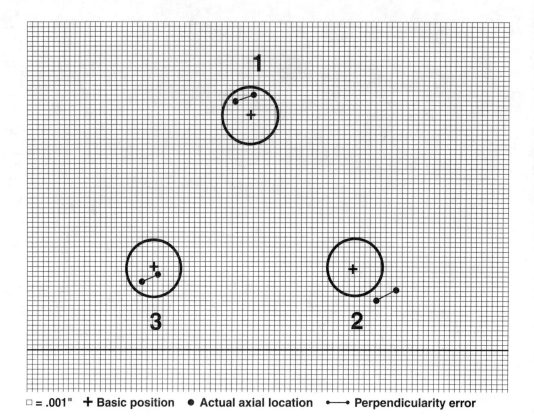

□ = .001"   **+** Basic position   ● Actual axial location   •——• Perpendicularity error

Figure 16.42 Tolerance zone overlay gage superimposed over the data graph using the .010″ - diameter zones.

Due to the nature of this callout, paper gaging is often not necessary. To inspect this part on an open surface-plate setup, the following equipment is necessary:

1. Surface plate
2. Dial indicator that discriminates to 10% or less of the position tolerance
3. Height gage to hold the dial indicator and make liner measurements
4. Two precision parallels (or right-angle plates) to contact the datums not contacted by the surface plate
5. The largest gage pins that can be inserted into the holes

STEP 1   Carefully rotate and slide the equipment onto the surface plate.

STEP 2   Insert the gage pins selected into the holes (one through all if possible).

STEP 3   Mount the primary datum against a parallel (or right-angle plate), the secondary datums on the surface plate, and the tertiary datum against another parallel (or right-angle plate), as shown in Figure 16.45. Clamp the part securely so that the setup can be rotated left 90° and the clamps will not interfere with surface-plate contact. Be careful not to damage the part during clamping.

Pattern Location, Feature to Feature, and Perpendicularity   265

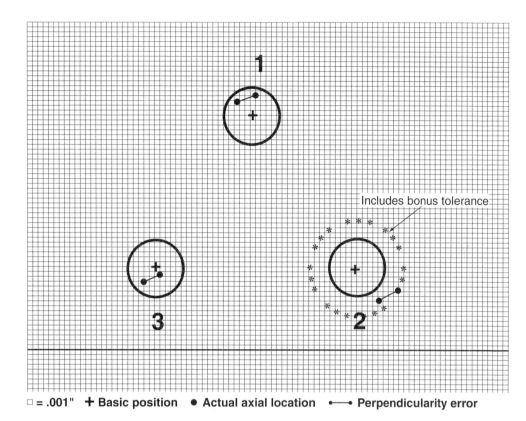

□ = .001"   + Basic position   ● Actual axial location   •—• Perpendicularity error

Figure 16.43 The overlay includes the applicable bonus tolerances for the final acceptance decision on the part.

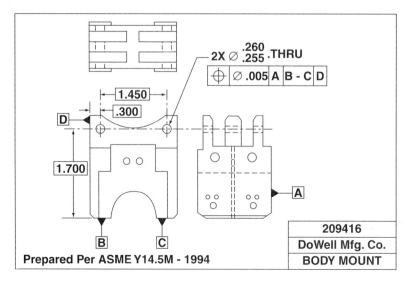

Figure 16.44 Drawing requiring a positional tolerance (for multiple controls).

266 Chapter 16 Position Tolerances

Figure 16.45 The part is mounted and clamped on all datums, the pins are inserted, and it is ready for inspection.

STEP 4 Measure the distance from the surface plate (datums *B* and *C*) to the top of the gage pin for each set of holes. Subtract one-half the pin size and record the results. Follow this procedure for *both ends* of the holes to evaluate the perpendicularity. (*Note:* An option is to inspect perpendicularity separately and relate the perpendicularity error to the tolerance zones. Refer to Figure 16.46.)

Figure 16.46 Making measurements on the long coordinate from multiple datums *B* and *C*.

Figure 16.47 The setup is rotated 90° and measurements are made on the other coordinates from datum *D* (at each end).

STEP 5 Rotate the setup 90° and measure the distance from datum *D* to the top of the pin in the first hole and from datum *D* to the top of the pin in the second hole, as shown in Figure 16.47. Follow this procedure for both ends of the holes to evaluate perpendicularity (or use the option previously stated).

STEP 6 Using the appropriate coordinates for each hole and distance, calculate the actual cylindrical tolerance zone for the position tolerance using the equation or table (See Figure 16.7) covered earlier in this chapter. Another option here is to prepare a paper gage. If each set of holes *at each end* is within the .005″ cylindrical tolerance zones, the part is acceptable. The observer should keep in mind that the cylindrical tolerance zones of .005″ diameter control pattern location, perpendicularity to datum *A*, and coaxiality of each set of three holes in line.

POSITION TOLERANCES: CYLINDRICAL PARTS, TERTIARY DATUM

Another example of position tolerancing on hole patterns that is often used is one that uses size feature datums and/or tertiary datums to *clock* the pattern. Clocking is the control of size features or patterns of features in rotation. The example shown in Figure 16.48 will be used to demonstrate both methods.

The tolerance zones for this requirement consists of *three cylindrical tolerance zones that are .010″ diameter only when the produced hole is at MMC size. These zones are equally spaced on the 2.100″ basic bolt circle diameter, perfectly perpendicular to datum plane B, and hole 1 is clocked at 60° basic from tertiary datum C. The imaginary bolt circle is perfectly concentric to datum axis A (when datum feature A is at MMC size).* The tolerance zones are shown in Figure 16.49.

268　Chapter 16　Position Tolerances

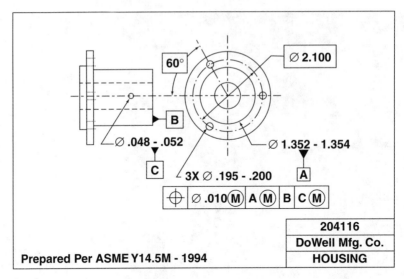

Figure 16.48 A drawing requirement for position of a hole pattern to size feature datums at MMC.

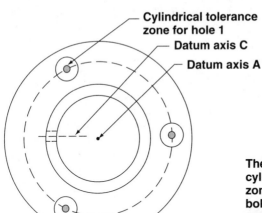

Tolerance Table

Actual Hole Size	Bonus Tolerance	Total Tolerance
.195	.000	.010
.196	.001	.011
.197	.002	.012
.198	.003	.013
.199	.004	.014
.200	.005	.015

The tolerance zones are .010" diameter cylinders (when the hole is at MMC). These zones are equally spaced on the 2.100" bolt circle, perpendicular to datum plane B, and hole 1 is clocked at a 60° angle from datum axis C. The bolt circle is centered on datum axis A (when axis A is at MMC). The axis of each hole must be within its cylindrical zone.

Additional tolerance is obtained in two ways:
1. If datum A is smaller than MMC size, an imaginary cylindrical zone is created around datum axis A to allow the center of the bolt circle to move (the pattern can shift from axis A).
2. If datum C is larger than MMC size, a total wide zone is created around datum C that allows the pattern to rotate further with regard to the clocking angle.

Figure 16.49 Tolerance zones for the part.

The drawing in Figure 16.48 requires that the three-hole pattern be located with respect to datum axis *A*, aligned on datum plane *B*, and clocked to datum axis *C* (where size feature datums are applicable at MMC). Pattern location to datum *A*, perpendicularity to datum *B*, and clocking to datum *C* must all be within the .010″ cylindrical tolerance zones. To inspect this part, the inspection setup would have to establish datum axis *A* by contacting datum diameter *A*.

Next, the setup must align the part on datum surface *B* and then index the part 60° and contact datum *C*. The inspection results, in this case, are best evaluated with a paper gage (or, if applicable, a functional gage). First, an example of a functional gage is shown in Figure 16.50. The functional gage design contacts all datums properly; then it has three virtual-size gage pins that must simultaneously be inserted into the three-hole pattern.

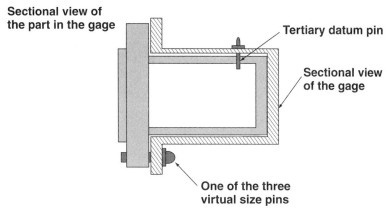

Note:
Three pins for the controlled holes are at the virtual size (.185" diameter).
The primary datum diameter in the gage is at MMC size (1.354" diameter).
The tertiary datum pin on the gage is at MMC size (.048" diameter).

Figure 16.50 Functional gage design for the position tolerance.

The use of the functional gage evaluates the worst-case conditions allowed by the position and size tolerances collectively *and physically rotates or shifts the part for best fit*. This rotation and/or shifting in an open setup can only be accomplished using a paper gage. It is also understood that direct measurement would have to be accomplished (and evaluated) if the position requirement were applicable at RFS (or LMC) instead of MMC.

To inspect the part in an open setup, the following equipment is needed:

1. Surface plate
2. Dial indicator that discriminates to 10% or less of the position tolerance
3. Height gage to hold the dial indicator and make linear measurements
4. Vee block (or rotary table)

5. Precision square, block, or other similar tool to contact datum B
6. Gage block set
7. The largest gage pins that can be inserted into the three controlled holes and the datum C hole

STEP 1 Calculate the linear basic dimensions (per the 2.100″ basic bolt circle shown) by solving the right triangle (or using the bolt circle chart in Appendix C). Remember that basic dimensions are perfect; therefore, they can be added, subtracted, or computed using other basic dimensions.

The following are solutions for the linear dimensions per the bolt circle chart in Appendix C. Refer to the drawing in Figure 16.48. Hole 1 has been identified as the hole that is in line with the small tertiary datum hole. The other two holes are numbered 2 and 3 going clockwise.

Hole No.	X Basic Coordinate (″)	Y Basic Coordinate (″)
1	1.05	0
2	.525	.9093
3	.525	.9093

STEP 2 Measure the size of the three controlled holes, the tertiary datum hole, and datum A diameter and record them on an inspection report for future use. Also, choose an order of the holes (for example, from the tertiary datum hole clockwise) and number each hole for consistent reference.

STEP 3 Carefully rotate and slide the inspection equipment onto the surface plate.

STEP 4 Insert the selected gage pins into the three controlled holes and the tertiary datum hole.

STEP 5 Mount the part datum A in a vee block (a rotary table or precision collect would be better here), and rotate the tertiary datum pin to where its *axis* is vertical, using a parallel to contact the pin in the tertiary datum. Once the part has been rotated, the two bottom holes should be parallel to the surface plate, as shown in Figure 16.51. The part should be clamped to the vee block using a standard vee-block clamp attachment (not shown in Figure 16.51).

STEP 6 Establish a zero point on the dial indicator by indicating at top dead center of datum A (while in the vee block); then subtract one-half of the size of datum A diameter, and move the height gage downward by that amount.

STEP 7 From that zero point, make linear measurements to the top of the gage pins for each hole; then subtract one-half the gage pin size, as shown in Figure 16.52. Record these measurements for future use according to the number of the holes and the direction (X or Y).

STEP 8 Rotate the setup 90° and repeat step 7.

Figure 16.51 The datums have been contacted, and the part has been located in the vee block for inspection.

It is now time, using this example, to discuss the difference between *bonus tolerances* and *additional tolerances*. First, bonus tolerances can be applied only to the location of the controlled holes *to one another* on the bolt circle (and then only when the holes are larger than their MMC size). Additional tolerance (depending on the type of datum feature) is an allowable shift, movement, or rotation of the *pattern* from the datum axis.

COMPOSITE POSITION TOLERANCES

Composite position tolerancing is a method that uses a double feature control frame for position of a pattern. The application is used when design requirements can allow more positional tolerance for pattern location than for the hole-to-hole requirement. In other words, the location of the pattern can have a looser tolerance than the features to each other. An example of a feature control frame for composite position tolerance is shown in Figure 16.53.

The drawing in Figure 16.54 is an example in which, composite tolerancing has been applied. The feature control frame states that the four-hole pattern must be within position tolerance cylindrical zones of .060″ (when the holes are at MMC) with respect to primary datum *A* secondary datum *B*, and tertiary datum *C*. The upper portion of the

272 Chapter 16 Position Tolerances

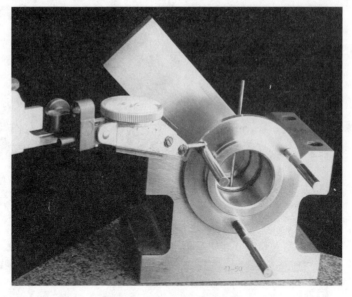

(a)

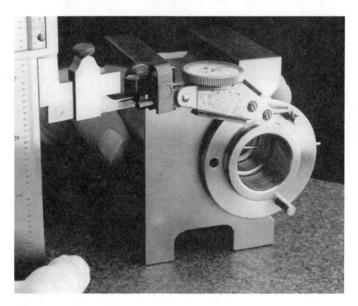

(b)

Figure 16.52 (a) Making the *Y* linear measurement from the zero point (datum axis) the axis of hole 1. (b) Making the *X* linear measurement for hole 1.

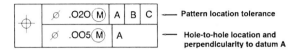

Figure 16.53 Feature control frame for a composite position tolerance.

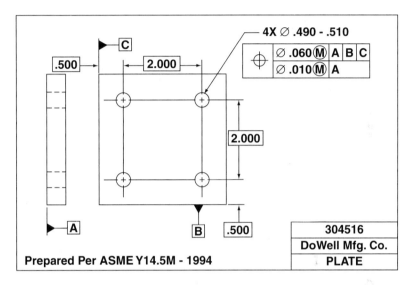

Figure 16.54 Drawing requirement for composite position tolerance on the three-hole part.

feature control frame also requires that the holes be perpendicular to datum plane *B* within the same .060″ cylindrical zones.

The lower portion of the feature control frame states that the pattern of holes to must be within position tolerance cylindrical zones of .010″ diameter (when the holes are at MMC size). The interpretation is that the tolerance zones locating the pattern are .060″ diameter *and* related to the datums directly. However, another set of tolerance zones (.010″ diameter) locate the holes to one another.

The inspection of this part will be a two-step process. The first method is to contact primary datums *A*, *B*, and *C*, and then to inspect the four holes directly from those datums for each coordinate location. Bonus tolerances would be allowed on the .060″ diameter zones if any hole is larger than MMC size. The next step requires that the holes be inspected *to one another* without contacting datums *A, B,* or *C* (with respect to the .010″) cylindrical zones at MMC. "Best fit" and bonus tolerances are allowed on the holes to one another. "Best fit," of course, would require a functional gage or a paper gage.

It is also important to note that the drawing in Figure 16.54 requires that the built-in perpendicularity tolerance shall be within the .030″ diameter zones, not the .010″ diameter zones. This requirement is so because datum *B* is specified in the upper portion of the feature control frame. If it had been specified after the .010″ tolerance, perpendicularity to datum *B* and hole-to-hole location would have to be within .010″ at MMC. There are other examples of composite position tolerancing, such as controlling the location of a pattern of holes that are in line with each other. A composite frame could locate these holes on the part with a looser tolerance and then control their coaxiality within a tighter tolerance (if design requirements dictated).

BIDIRECTIONAL POSITION TOLERANCES

A relatively new application for position tolerances (found in the 1982 revision of the standard) is bidirectional position tolerancing. Bidirectional position tolerancing, when design requirements dictate, allows more position tolerance in one direction than in another for one or more controlled feature(s). Consider, for the example, the drawing in Figure 16.55, which shows how bidirectional position tolerances are specified.

The feature control frames state that the centerplane of the slot shall be within a total wide position tolerance zone of .030″ from datum plane B (when the slot width is at MMC size) and a total wide tolerance zone of .005″ from datum A (when the slot height is at MMC size).

The drawing in Figure 16.55 shows that, from datum *A*, the position tolerance is .005″ *total wide zone*, and from datum *B*, the position tolerance is .030″ *total wide zone*. The centerplane of the slot must lie within each of the zones at the same time. These total wide zones, in effect, generate a *rectangular tolerance zone*, as shown in Figure 16.56. Unlike old plus and minus coordinate tolerancing, these total wide zones are specified at MMC and bonus tolerance can be achieved.

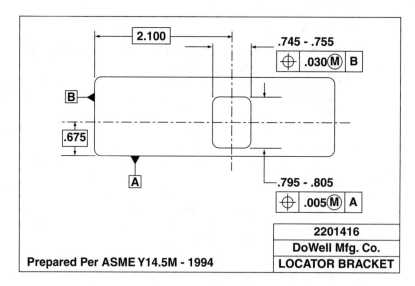

Figure 16.55 Drawing requirement for bidirectional position tolerance.

Bidirectional Position Tolerances   **275**

Bonus Tolerance Table

.745" - .755" Dimension			.795" - .805" Dimension		
Actual Hole Size	Bonus Tolerance	Total Tolerance	Actual Hole Size	Bonus Tolerance	Total Tolerance
.745	.000	.030	.795	.000	.005
.746	.001	.031	.796	.001	.006
.747	.002	.032	.797	.002	.007
.748	.003	.033	.798	.003	.008
.749	.004	.034	.799	.004	.009
.750	.005	.035	.800	.005	.010
.751	.006	.036	.801	.006	.011
.752	.007	.037	.802	.007	.012
.753	.008	.038	.803	.008	.013
.754	.009	.039	.804	.009	.014
.755	.010	.040	.805	.010	.015

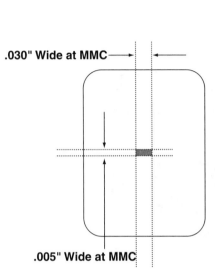

.030" Wide at MMC

.005" Wide at MMC

The tolerance zone, in effect, is a rectangular-shaped zone because there are two sets of parallel planes intersecting each other. Since the tolerance zones are called out at MMC, bonus tolerance can be achieved if either the width or height (or both) are beyond MMC size.

The rectangular tolerance zone is located by the 2.100" and .675" basic dimensions. More tolerance is allowed in the direction from datum B (.030" at MMC) than from datum A (.005" at MMC). The bonus tolerance can only be used directionally.

Figure 16.56 Tolerance zone for the part.

The inspection method for this part requires the observer to make directional coordinate measurements for each direction *locating only on the one specified datum for that direction*. Each specified datum must be treated as a primary datum. Then, allowing for any bonus tolerance, accept the part if each coordinate centerplane is within its directional total wide tolerance zone. An important thing to remember is that each direction has a primary datum that must be located on its three (or more) highest points. It is also important to note that the controlled feature could also be a hole and that the tolerance zone is still a *total wide zone in each direction*. The equipment required for open surface-plate inspection is as follows:

1. Surface plate
2. Dial indicator that discriminates to 10% or less of the *tightest* positional tolerance
3. Height gage to hold the dial indicator and to make linear measurements
4. Set of adjustable parallels
5. Dial calipers (or another appropriate instrument to measure the slot size)

STEP 1 Measure the width of the slot in each direction and record the information for future use on a piece of paper (or inspection report).

STEP 2 Find an adjustable parallel with a range that can be inserted into the slot; then insert, spread, and lock the parallel in a manner that covers the flat area of the slot (and avoids the corner radii), as shown in Figure 16.57.

Figure 16.57 An adjustable parallel is inserted and locked into the slot.

STEP 3 Place the part on the surface plate contacting datum *A* at its three (or more) highest points, as shown in Figure 16.58. This setup will compare the centerplane of the part to the single .675″ basic coordinate. Measure the distance from the surface plate to the top of the adjustable parallel, and subtract one-half the slot size from that measurement. Record this value on a piece of paper (or inspection report) as the actual value from datum *A*.

STEP 4 If necessary, find another adjustable parallel to fit into the slot in the other direction (in this case, the same parallel was used). Insert and lock the parallel in place in the same manner as before.

STEP 5 Locate datum *B* on the surface plate and measure the distance from the plate to the top of the parallel once again, as shown in Figure 16.59. Subtract one-half the slot size from that measurement and record the value on a piece of paper (or inspection report). This setup will compare the centerplane location of the slot to the 2.100″ basic coordinate.

Bidirectional Position Tolerances 277

Figure 16.58 The first coordinate is measured using the adjustable parallel.

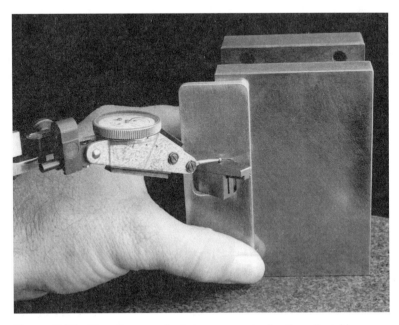

Figure 16.59 The other coordinate is measured using the adjustable parallel.

STEP 6 Evaluate the results. In this case, a paper gage is not necessary, but an inspection report, such as the one in Figure 16.60, will help evaluate the part. The inspection report shows the necessary information and the results of inspection. This part, for example, is not acceptable because the slot is out of location tolerance from datum *A*.

Datum *A* Coordinate Measurements

Basic Dimension	Actual Dimension	Coordinate Difference	Actual Slot Size	Applicable Bonus Tolerance	Total Position Tolerance	Accept or Reject?
.675″	.681″	+.006″	.800″	.005″	.010″ Total (±.005″)	**Reject**

Datum *B* Coordinate Measurements

Basic Dimension	Actual Dimension	Coordinate Difference	Actual Slot Size	Applicable Bonus Tolerance	Total Position Tolerance	Accept or Reject?
2.100″	2.110″	+.010″	.745″	.000″	.030″ Total (±.015″)	**Accept**

Figure 16.60 The inspection report shows that the part is out of tolerance at the .675″ coordinate.

Bidirectional position tolerancing (when specified at MMC) allows for functional gage design and use when the most practical decision is gaging. The functional gage shown in Figure 16.61 is a relatively simple design. The gage must be designed to contact each primary datum one at a time. This can be accomplished with removable faces or two gages. The gage must also have a rectangular pin that has both different virtual conditions, as shown in the figure. The rectangular pin should be removable so that the primary datum can first be contacted; the pin can then be inserted. (*Note:* If the feature being located were a hole, instead of a slot, the gage pin would be a diamond-shaped pin with radii at the corners. This pin would also encompass both virtual sizes for the bidirectional position tolerance applied to the hole.)

PROJECTED TOLERANCE ZONES

Projected tolerance zones can be applied to other than position tolerances, but position will be used for this example. The typical application for projected tolerance zones is for mating parts that have the *fixed fastener case*. Fixed fasteners are examples of mating parts fastened together with threaded fasteners, press fit pins (or studs), rivets, or other fasteners that are in the *fixed* condition in assembly.

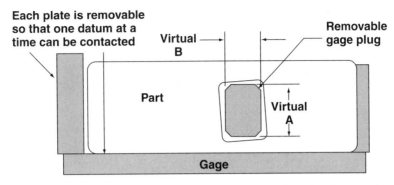

Figure 16.61 Functional gage for bidirectional position tolerance at MMC.

Floating fasteners are the opposite of fixed fasteners in that there is still clearance between the mating parts where the fastener goes into the part. For example, two parts that have clearance holes in them and a bolt that goes through both holes is a floating fastener case. The bolt is not restricted when passing through the part. Projected tolerance zones are usually not necessary in the floating fastener case.

Figure 16.62 shows an example of a fixed fastener case that tends to explain why projected tolerance zones are necessary. In this figure, we can see that the press fit studs are fixed in their holes in part B. The lean of the holes in part B (allowed by the typical position tolerance zone) causes an extreme virtual condition of the studs. This *projection* of the studs can cause problems in attempting to assemble the mating part A.

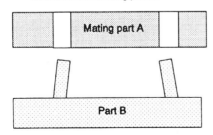

Figure 16.62 Example of an application for projected tolerance zones.

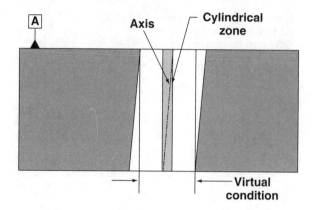

Figure 16.63 Typical position tolerance zone description for a hole.

For a comparison, refer to Figures 16.63 and 16.64. Figure 16.63 shows, a typical position tolerance zone. This zone controls only the location and perpendicularity of the axis of the hole, not the stud that will be pressed into the hole in subassembly. Figure 16.64 provides a projected tolerance zone that begins at the interface of the mating parts (datum *A*) and extends through the studs for a distance equal to the maximum thickness of the mating part. Therefore, the virtual condition of the studs at subassembly is controlled by virtue of the project tolerance zone for the holes.

An example of an assembly that requires projected position tolerance zones is shown in Figure 16.65. In this case, the top part has holes that will be threaded at a fu-

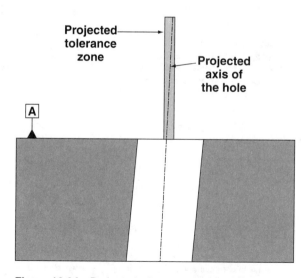

Figure 16.64 Projected tolerance zone for a hole.

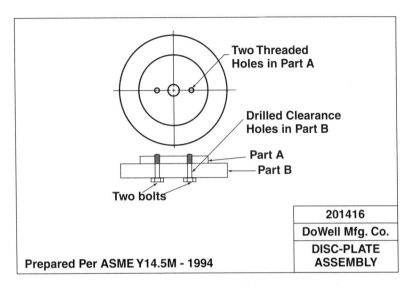

Figure 16.65 Assembly application for projected tolerance zones.

ture operation; then the top will be assembled to the bottom part. The position tolerance for the threaded part is shown in Figure 16.66 with projected tolerance zones. The feature control frame states that the axis of each hole must lie within a projected cylindrical tolerance zone of .005″ diameter (when the hole is at MMC size) with respect to primary datum plane A. Perpendicularity of each hole to datum plane A is also required within the same cylindrical tolerance zones. The projected tolerance zone requirement on the drawing is specified with a circled P, and the minimum height of the projection is given inside the box under the feature control frame. The *chain line* is used to show the direction of the projected tolerance zone (since the holes are through the part). If there are blind holes in the part, the direction of the projected tolerance zone is automatically upward.

The inspection method for this type of part has already been covered in earlier examples in this chapter. The primary difference, in this case, is that all position and perpendicularity measurements will be made on the extended portion of the gage pins only. No measurements should include the opposite end of the hole.

The projected tolerance zone begins at datum A and extends upward. Due to this extension, all the measurements the observer needs to make are at each end of the exposed pin. Functional gage designs (for projected tolerance zones) have to serve the same purpose. They should be designed so that the virtual condition of the projected axis of the hole is gaged, not the axis of the hole itself. An example functional gage for this part is shown in Figure 16.67. Projected tolerance zones are a very specific type of tolerance zone that allows the designer to accommodate the mating part assembly early in the manufacturing of the detail parts.

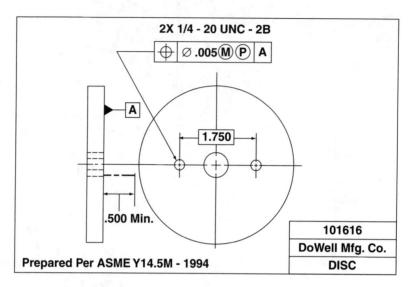

Figure 16.66 Detail part drawing requirement for a position tolerance with projected tolerance zones.

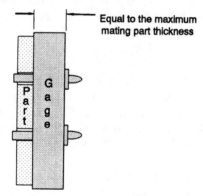

Figure 16.67 Functional gage design example for projected tolerance zones.

ZERO POSITION TOLERANCE AT MMC

Zero position tolerance at MMC is a concept that has been widely used to control the location of size features and size feature datums. This application is often used on size feature datums to qualify the datum.

The datum–virtual condition rule of the standard relates to size feature datums and the fact that they may have a virtual condition even when they have been modified at MMC. Zero tolerance at MMC, when applied to the datum feature controlled, accounts for this by making the virtual size (worst case) of the datum equal to the MMC size.

The application of zero tolerance at MMC does not mean that there is no tolerance. It means that the position tolerance is zero *only when the feature being located with po-*

sition tolerance is produced at MMC size. All the position tolerance allowed for the feature is dictated by the amount of bonus tolerance allowed.

Some of the advantages of using zero tolerance at MMC are as follows:

1. The virtual size of the feature is controlled (equal to the MMC size).
2. It eliminates the need to apply the datum–virtual condition rule
3. When functional gages are used, the virtual size member of the gage acts as a worst case *and* a "go" gage for size.
4. More tolerance for location can be allowed without impairing the function of the part (in certain cases).

Inspection of parts with zero tolerance at MMC is no different than other position tolerances already covered in this chapter. The exception is interpreting the amount of tolerance allowed. The tolerance allowed is equal to the applicable bonus tolerance (or the difference between the actual size of the controlled feature and the MMC size of that feature).

An example of zero position tolerance at MMC is shown in Figure 16.68. The feature control frame states that the holes must be in perfect position (with respect to datums *A, B,* and *D*) if they are produced at their MMC size. In this example, if the holes are produced at MMC size (.500″), they must be in perfect position (or exactly on their basic location). The tolerance zone allowed for a hole that is .501″ diameter is a .001″ cylindrical zone (because there would be .001″ bonus tolerance allowed).

The tolerance allowed for a hole that is .503″ diameter is a .003″ cylindrical zone (because there would be .003″ bonus tolerance allowed). If the holes are produced at

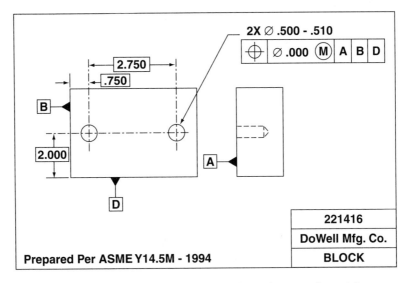

Figure 16.68 Drawing requirement for a position tolerance of zero tolerance at MMC.

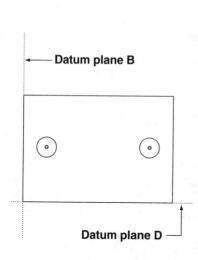

Figure 16.69 Tolerance zones for the part depend on hole sizes.

LMC size (.505″ diameter), the position tolerance is equal to all the bonus tolerance allowed (.005″ cylindrical zone). The tolerance zones and a table of example tolerances are shown in Figure 16.69.

POSITION TOLERANCE AT LMC

When position tolerances are applied at the LMC size of the controlled feature, a bonus tolerance is allowed. This bonus tolerance, however, works differently than MMC. In these applications, the tolerance zone applies when the controlled feature is at LMC size, and bonus tolerances are allowed as the feature departs from LMC size toward MMC size. To understand how position tolerances at LMC work, the following examples are given.

Example 1

Round Boss with a Hole in It (coaxially): When position tolerances at LMC are applied to locate the axis of the round boss and the axis of the hole, they automatically control the minimum wall thickness and allow maximum tolerances for the location of both features. As the outside diameter of the round boss departs from LMC size (its smallest size), more position error is allowed. As the hole departs from LMC (its largest allowable size), the hole is allowed more position error. In either case (or both cases), the minimum wall thickness, as designed, is guaranteed.

Example 2

Round mounting pads on a casting, when positioned at LMC, guarantee that the locator pins on a machine will hit the pads even though they are out of position. This is so because the pads must be in position when at LMC (their smallest diameter), and if they are larger than LMC (toward MMC size), they can be out of position further. There will still be enough surface of the pad to come in contact with machine locator pins.

The drawing in Figure 16.70 shows an example of position tolerancing at LMC. The inspection methods for this part have already been covered in this chapter, but the observer must know that the feature control frame states a .005″ cylindrical zone of position tolerance only when the holes are produced at LMC size (.510″ diameter). If the holes are smaller than LMC, a bonus tolerance can be added to the stated .005″ tolerance.

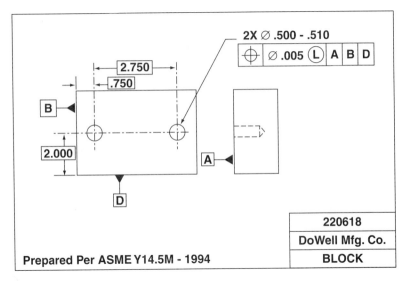

Figure 16.70 Drawing requirement for position tolerance at LMC.

For example, if the actual size of a hole is .508″ diameter, the total position tolerance allowed for that hole would be .007″ cylindrical tolerance zone (.005″ stated tolerance plus .002″ bonus tolerance), as shown in Figure 16.71.

Position: Two Single Segment

The two single segment is a new application for position tolerancing (shown in the 1994 standard). The application is similar to composite position tolerancing, except that two datums are repeated in the lower portion of the feature control frame. The drawing in Figure 16.72 shows that the outside edge datum feature *B* has been repeated

Chapter 16 Position Tolerances

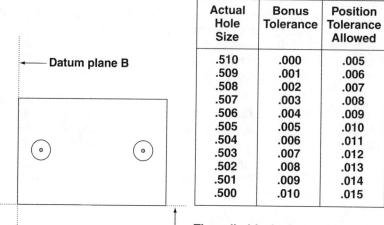

Actual Hole Size	Bonus Tolerance	Position Tolerance Allowed
.510	.000	.005
.509	.001	.006
.508	.002	.007
.507	.003	.008
.506	.004	.009
.505	.005	.010
.504	.006	.011
.503	.007	.012
.502	.008	.013
.501	.009	.014
.500	.010	.015

The cylindrical tolerance zone for each hole is perpendicular to datum plane A and located by basic dimensions from datum planes B and D.

If the holes are produced at LMC size, they must be in position within a .005" diameter cylindrical zone. If either hole is produced smaller than LMC size, a bonus position tolerance is allowed that is equal to the amount of departure from LMC size. Refer to the above table for examples.

Figure 16.71 Tolerance zones for the part.

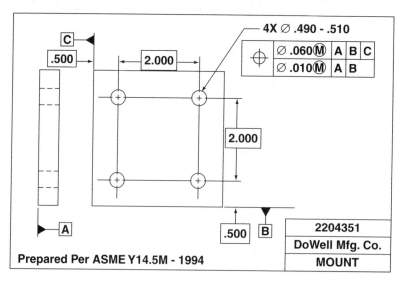

Figure 16.72 Two single-segment position tolerance.

in the lower frame. This is because, in the assembly, datum feature *B* has no clearance. The .010″ cylindrical tolerance zones are located to each other, perpendicular to datum *A*, and fixed in location to datum *B*. In this case, the .060″ cylindrical tolerance zones allow the designer to share the clearance at datum *C* with manufacturing in terms of a looser tolerance for location from that datum.

The tolerance zones for the two single-segment position tolerance are shown in Figure 16.73. It is important to understand that the only requirement is that all four axes of the holes must fall inside, or on the edge of, both of the tolerance zones at the same time. The four .010″ zones can float in the direction of datum *C*, but are rigidly located from datum *B*.

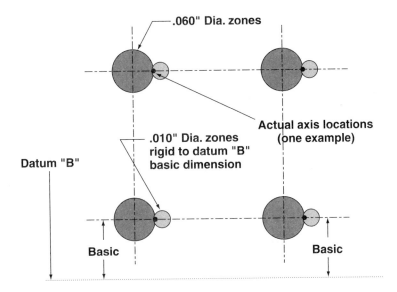

Figure 16.73 The tolerance zones for two single-segment position tolerance.

BOUNDARY POSITION TOLERANCING

Boundary position tolerance is another new application for position that was developed in the 1994 standard. Boundary position tolerance is a special application. It is used when the actual surfaces of the size feature must be contained to the point where they cannot fall inside an inner boundary that is generated by the size and position tolerance. The significant difference between normal position tolerancing and the boundary method is that the word BOUNDARY is specified under the feature control frame. Figure 16.74 shows a square slot that has been positioned using the boundary control method.

288   Chapter 16   Position Tolerances

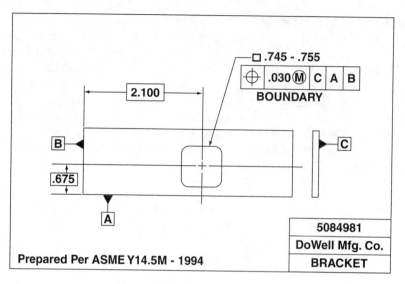

Figure 16.74   Boundary position tolerance.

The imaginary tolerance zone for this boundary position tolerance (as shown in Figure 16.75) is a perfect square that is located by the basic dimensions. The size of the square zone is equal to the MMC size of the square feature minus the position tolerance of .030″. Inspection of this part requires the following steps:

1. Calculate the inner boundary zone. The zone is calculated by subtracting the position tolerance (.030″) from the MMC of the square feature (.745″). The zone therefore is a .715″ square.
2. Locate datum A on the surface plate. Using the .675 basic dimension, subtract one-half the tolerance zone (.3575″). This is equal to .3175″.
3. Indicate the bottom surface of the slot. There should be no readings higher than .3175″ from datum A.
4. Next, add one-half the .715″ zone to .675″ where the sum is 1.0325″. Invert the dial indicator, and set zero at 1.0325″ from datum A. Indicate the upper wall of the slot. There should be no readings below zero at the upper wall.
5. Next, using datum B and the 2.100 basic dimension, repeat the aforementioned procedure.

Position Tolerance: Separate Requirements

There are times when two or more separate patterns of features share a common basic bolt circle, as shown in Figure 16.76. This drawing shows a two-hole pattern (the .185 and .195 holes) and another two-hole pattern (the .495 and .505 holes) on the same ba-

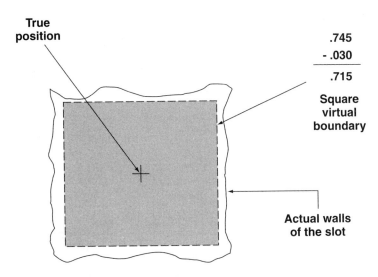

Figure 16.75 The tolerance zone for boundary position tolerance.

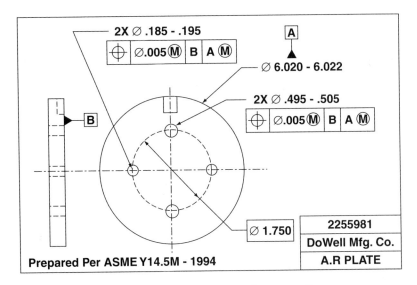

Figure 16.76 The two sets of holes are treated as one pattern.

sic bolt circle. In these cases, it is understood by the requirements of the standard that these two sets of holes must be treated as one four-hole pattern with respect to the position tolerances. This is true even though there are two separate position callouts.

If the two sets of hole patterns were independent in final assembly, then they are not directly related to each other. If that is the case, they are separate requirements. In order

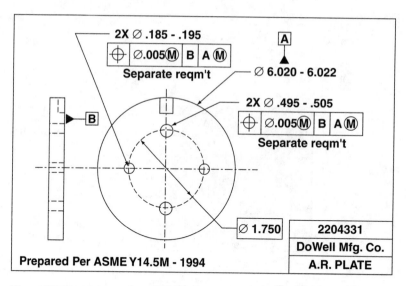

Figure 16.77 The two sets of two-hole patterns now have separate requirements.

to clearly specify a separate requirement, the words SEPARATE REQM'T must be specified under each of the position feature control frames (as shown in Figure 16.77).

To inspect the part in Figure 16.76, methods already covered earlier in this book are used, but the two sets of two-hole patterns are inspected together, and all four holes can share their clearance in terms of "best fit." To inspect the part in Figure 16.77 for separate requirements, each set of two-hole patterns is evaluated separately.

Position Tolerance: Conical Tolerance Zone

Another new position application in the 1994 standard involves positioning size features using a conical-shaped tolerance zone (as opposed to the traditional cylindrical zone). Consider the drawing shown in Figure 16.78. In order to specify a conical zone for the location of the axis of this hole, the designer must label specific surfaces at each end of the hole and then use two different position callouts for the same hole. Figure 16.78 shows that one of the surfaces has been labeled surface C and the other has been labeled surface D. Notice that surface D is adjacent to a small slot in the bottom of the part.

The drawing also shows two different feature control frames for the hole. One frame allows a .020″ diameter zone at surface C; the other requires a tighter diameter zone of .005″ at surface D. Since the zone is .005″ diameter at one end and .020″ diameter at the other end, the tolerance zone is the shape of a cone (as shown in Figure 16.79). Inspection of this part is to inspect the location of the axis of the hole, but allow .005″ at one end and .020″ at the other.

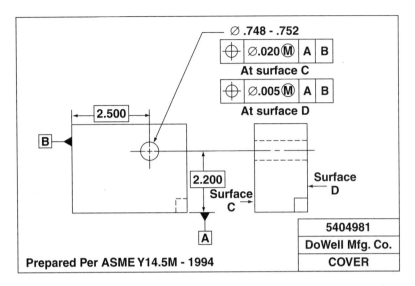

Figure 16.78 A drawing for conical position tolerance.

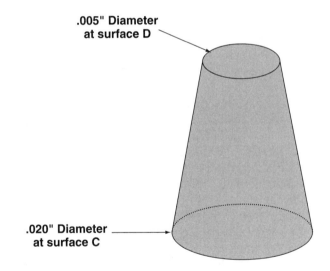

Figure 16.79 The conical position tolerance zone.

Position Tolerance at LMC

In some applications restrictions require a specific kind of tolerancing; the part in Figure 16.80 is one example. This part is under stresses in the final assembly where the minimum wall thickness from the edge of the holes to the outside diameter of the part must be controlled. Typically, the position tolerance might be applied using the

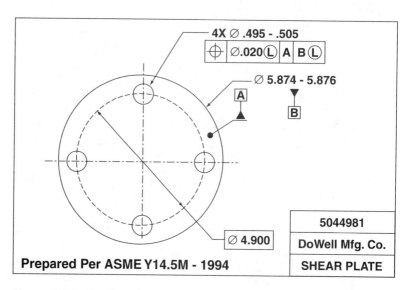

Figure 16.80 Position tolerance stated at LMC.

regardless of feature size (RFS) modifier so that the resulting wall thickness would be controlled. Note that the maximum material condition (MMC) modifier, in this case, will not work because of the bonus position tolerance that is allowed as the holes are produced larger than MMC size.

On the other hand, it is always good to be able to allow bonus tolerance for position of the holes wherever practical. This is a classic example of how position tolerance at LMC works. The drawing in Figure 16.80 specifies that the hole sizes must be between .495″ and .505″ in diameter. Note that the LMC of the holes is the largest hole size (.505″). It also specifies that each hole has a position tolerance of .020″ cylindrical zone when the hole is at LMC size. It then specifies that the datum for location of the hole patter is the outside diameter (datum *B*) when it is at LMC size (5.874).

Position tolerance at LMC allows bonus position tolerance as the feature being controlled departs from LMC size toward MMC size. The table in Figure 16.81 shows the position tolerance allowed for each hole within the pattern at given actual-size intervals of .001″ (for example only). Because of the way position tolerance at LMC works, the minimum wall thickness is guaranteed. As the hole gets smaller, it increases the minimum wall thickness. Since a smaller hole provides greater wall thickness, the axis of the hole is allowed a larger position tolerance zone. The result is that the edge of the controlled hole goes back to the minimum wall thickness that has been designed.

Now, let's focus on datum *B*, which has also been specified to apply at LMC size. Specifying datum *B* at LMC size, means that, when datum *B* diameter is produced at LMC size (5.874″), the pattern of holes must be centered on axis *B*. As the outside diameter of datum *B* departs from LMC size toward MMC size, a shift in the center of the four-hole pattern is allowed. LMC on the datum also is intended to allow "additional" tolerance for pattern location as long as datum *B* is larger than LMC size. A larger datum *B* diameter increases the minimum wall thickness; therefore the pattern

Actual Hole Size	Stated Position Tolerance	Bonus Tolerance	Total Position Tolerance Allowed
.505 (LMC)	.020	.000	.020
.504	.020	.001	.021
.503	.020	.002	.022
.502	.020	.003	.023
.501	.020	.004	.024
.500	.020	.005	.025
.499	.020	.006	.026
.498	.020	.007	.027
.497	.020	.008	.028
.496	.020	.009	.029
.495	.020	.010	.030

Figure 16.81 Table of bonus position tolerance based on LMC size.

of holes can shift by that amount, resulting in a shift back to the minimum wall thickness designed.

COORDINATE MEASURING MACHINES

The examples shown thus far in this chapter have to do with open-setup measurements of position tolerances on a surface plate. When coordinate measuring machines (CMMs) are used, the setup is not very different from those already shown. It is always best to use surface-plate accessories (such as right-angle plate and precision parallels) on the surface plate of the CMM to contact datums properly, and it is still best to insert the largest true gage pin in holes for a more functional inspection (especially in cases where the hole-making process produces out-of-round holes). CMMs come with various types of probes from the electronic probe to fixed probes. The electronic probes are often used to touch various points on a surface or inside a hole to establish the datum plane or datum axis, respectively. With most CMMs, when the probe touches the part, a beeping sound is heard.

The practice of beeping a surface to establish a datum plane is often error-prone because the observer does not know where the highest points of the surface are located. Hence, the observer cannot locate the datum at the highest points.

The same holds true for beeping the inside of a hole to measure the position of its axis, as discussed earlier in this chapter. When holes are out of round, tapered, or the like (a result of several hole-making processes), the observer never really knows whether the axis has been established or not. When surface-plate accessories are used, the functional datums and axes are sure to be contacted. To contact a primary datum plane, the observer simply locates the part on the CMM's surface plate.

To contact a secondary datum plane, the observer can use a right-angle plate, beep the right-angle plate at any two points, and then bring the secondary part datum surface in contact with the right-angle plate. To establish a tertiary datum, the observer can use another right-angle plate that is 90° from the first one, beep this plate in one location, and then bring the tertiary datum feature against the right-angle plate.

294   Chapter 16   Position Tolerances

Many companies use fixtures for the CMM when production quantities warrant the expense. In such cases, observers can beep the fixture surfaces (three, two, and one point, respectively) and then mount the part in the fixture for inspection. All the example setups shown in this chapter are intended to show the way the part should be set up for inspection. Using a CMM just makes the setup and inspection faster.

The use of a CMM does not mean that the setup can all be accomplished using the beep probe and probing random points to establish datums or to measure axial locations of holes. When using CMMs, the observer must also recognize that the CMM is a *coordinate measuring machine*, and certain position tolerances require graphical inspection analysis to make the correct acceptance decision.

CMMs (not equipped with software) do not assist the observer in rotating data for best fit; therefore, the observer should still use graphical inspection analysis on the CMM results. There are computer-assisted and computer-driven CMMs that can perform graphical inspection analysis if they are equipped with the software for this method.

To summarize, if observers are able to use a CMM to perform the inspection of position tolerances faster and more efficiently, they should still take the time to set up the inspection properly and to properly analyze the results.

Example 16.3

An earlier part (Figure 16.9) that was inspected in open setup could also be inspected on a CMM using the same surface-plate accessories as before. For CMM inspection, the precision parallels should still be used to provide simulated datums to establish the datum reference frame for the part. In certain cases, depending on the accuracy of the process that produced the holes, the use of gage pins to find the axes of the holes may not be required. For example, if the holes were drilled and reamed in the part, the circularity error may not be significant enough to worry about axial measurement error. In cases like this one and in many others, the CMM probe can be used directly inside the hole to find the axis. The observer has to use good judgment and knowledge of the process when making the decision to use pins or not.

The CMM used in this example is computer-driven. The computer has appropriate software to evaluate geometric relationships. Also included is an electronic beep probe used to probe the surfaces of the holes. The following steps should be used to inspect this part on the CMM:

STEP 1   Carefully slide the precision parallels (or right-angle plates) onto the CMM surface plate.

STEP 2   To establish the primary datum plane, it is most effective and functional to first beep the CMM surface plate in three places so that the computer generates a primary datum plane from a true plane. Figure 16.82 shows the probe beeping the surface in one of three places.

STEP 3   To establish the secondary alignment datum for the part, the observer should beep one of the precision parallels in two places, as shown in Figure 16.83. In this manner, the secondary datum feature of the part can be properly established.

STEP 4   Finally, as shown in Figure 16.84, the other precision parallel is placed at 90° from the first parallel (to contact the tertiary datum surface), and this parallel is probed in one place to complete the datum reference frame for

Figure 16.82 The CMM probe is establishing the primary datum plane for the part. (Courtesy of Renishaw® and Giddings & Lewis®).

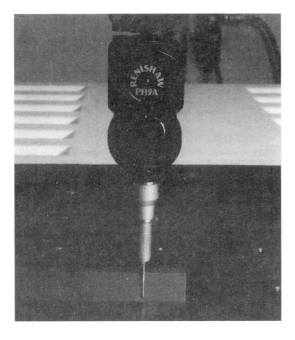

Figure 16.83 The CMM probe is establishing the secondary datum plane for the part. (Courtesy of Renishaw® and Giddings & Lewis®).

296 Chapter 16 Position Tolerances

the part. Depending on the CMM and software, it will be necessary to enter basic dimensions (X and Y) for the holes into the computer at this, or at an earlier step. (*Note*: At this point, the parallels are in position, and the datum planes for the part have been established for CMM inspection.)

STEP 5 Mount the part in the setup with the primary datum on the CMM surface plate, the secondary datum against the first precision parallel, and the tertiary datum (end surface) against the second parallel. From here the CMM probe can be used to inspect the position of every hole in the part without further movement of the part or setup.

STEP 6 The CMM probe is now carefully inserted into the hole and the hole is probed in at least three places (120° apart), as shown in Figure 16.85. For best results, one of the probed points is at the bottom of the hole, the next is 120° and at the middle of the hole, and the last point is 120° rotated and at the top of the hole. The result of this probing is a scope of points that can be used to generate an axis. The observer should probe all the holes in this manner.

Example 16.4
Another example for CMM inspection is the part in Figure 16.22, discussed earlier in this chapter. The CMM could also be used to measure the four-hole position tolerance, as shown in Figure 16.86. In this setup, the probe has already beeped the CMM surface plate in at least three places to establish the primary datum plane for the part before the part was placed on the CMM surface plate. Once the datum plane has been established (in the computer) and the basic coordinates have been entered, all four holes can be probed for inspection. Keep in mind that the

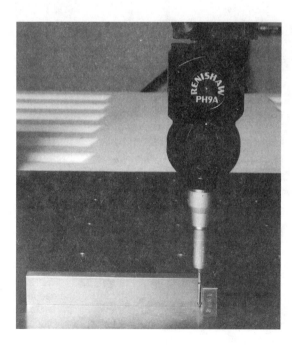

Figure 16.84 The CMM probe is establishing the tertiary datum plane for the part. (Courtesy of Renishaw® and Giddings & Lewis®.)

Figure 16.85 Measuring the hole locations on the CMM. (Courtesy of Renishaw® and Giddings & Lewis®.)

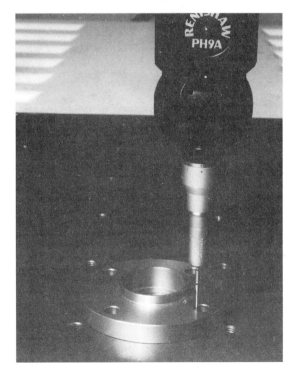

Figure 16.86 Measuring the four hole locations on the CMM. (Courtesy of Renishaw® and Giddings & Lewis®.)

probing of each hole should occur at least three points, 120° apart, from top to bottom of the hole. This is especially true in this case because perpendicularity is a built-in requirement of this position tolerance application.

Summary

The CMM, properly used, can be a faster, more accurate, and more efficient method for inspection of position tolerances depending on many circumstances, such as cost and amount of inspection required. CMMs that are computer-driven (or at least computer-assisted) can be equipped with geometric tolerancing software that will interpret the results of position tolerances more efficiently.

REVIEW QUESTIONS

1. Position tolerances are often used to locate single surfaces. *True* or *False?*
   Match the control descriptions on the left to the correct tolerance zone shapes on the right:

   Position Control Description
   2. Axis of a hole
   3. Centerplane of a square slot
   4. Axis of a shaft
   5. Bidirectional hole coordinate

   Tolerance Zone
   A. Two parallel planes
   B. Cylinder
   C. Tubular
   D. Two parallel lines
   E. Two concentric circles
   F. Rectangular zone

6. Bonus position tolerances can be applied, depending on actual feature sizes, when position tolerances are modified with either MMC or LMC. *True* or *False?*
7. The best fit of a hole pattern cannot be seen using simple coordinate measurements alone. *True* or *False?*
8. Position tolerances must always be specified with respect to datums. A position tolerance for a hole pattern that does not specify a datum reference in the feature control frame still has datums. They are called _____ datums.
9. According to general rule 2, a position tolerance with no modifier specified is understood to apply RFS. *True* or *False?*
10. What is the only position tolerance modifier that allows the use of functional gaging for inspection of position tolerances?
11. When position tolerance is applied to a hole, does the virtual condition result of that tolerance effectively make the hole bigger or smaller?
12. Position is often used to control the coaxiality of nonrotating parts. In these cases, the position tolerance could be called out using MMC or RFS modifiers. *True* or *False?*
13. In a composite position tolerance feature control frame, the tolerance value specified in the upper block specifies the tolerance zone for (a) feature to feature, (b) perpendicularity, (c) pattern location, (d) angularity.
14. There are two very effective methods for analyzing position tolerances for best-fit requirements. One of them is _____ analysis.
15. When applying position tolerances on the basis of zero tolerance at MMC, the entire tolerance zone for the location of the feature depends on the amount of _____ tolerance that is allowed.

17

Symmetry

INTRODUCTION AND APPLICATIONS

KEY FACTS

1. The symmetry symbol was deleted in the 1982 revision of the standard, and then reinstated in the 1994 revision.
2. Symmetry applications are, more often than not, for cosmetic (appearance) purposes rather than functional purposes.
3. Functional applications for symmetry involve features that must be equal in distance from a datum reference or the components will not assemble.
4. Symmetry requires at least one datum reference.
5. Symmetry tolerance is applied only on a regardless of feature size (RFS) basis. Modifiers such as maximum material condition (MMC) or least material condition (LMC) do not apply.
6. Since symmetry is always applied on an RFS basis, functional gages cannot be used for inspection. Symmetry should be measured using differential measurements in order to find the location of the axis or centerplane of the feature.

Symmetry is a tolerance that requires a size feature to be evenly centered (equidistant) between two other features or evenly centered on another size feature. For example, in the drawing shown in Figure 17.1, the feature control frame requires that the centerplane of the middle tab be symmetrical (centered) on the width of the part (datum *A*, RFS) within a tolerance of .004″ (RFS).

Although the size of the middle tab and datum *A* may vary within their size tolerances, the tab must be symmetrical within the stated .004″ tolerance zone. The tolerance zone for this relationship (as shown in Figure 17.2) is *two parallel planes .004″ apart and centered (or equally disposed) on the centerplane of datum feature A.*

Chapter 17 Symmetry

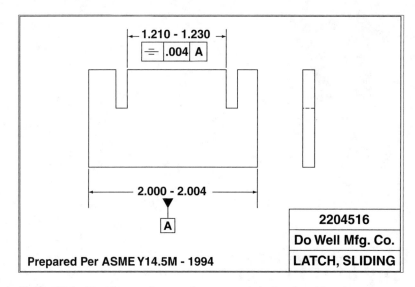

Figure 17.1 Drawing requirement for symmetry (using the old symmetry symbol).

The tolerance zone is two parallel planes that are .004" apart (regardless of feature size). The tolerance zone planes are equally disposed around datum centerplane A. The centerplane of the controlled feature must lie within the tolerance zone.

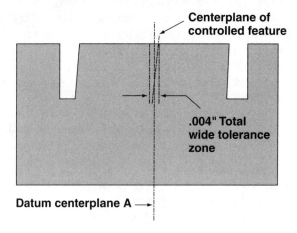

Figure 17.2 Tolerance zone for the part.

One of the fastest methods of inspecting this part is with simple differential measurements. There are many ways to locate the part in a surface-plate setup. The following method is but one example. The equipment required to inspect this part in an open surface-plate setup is as follows:

1. Surface plate
2. Dial indicator that discriminates to 10% or less of the position (symmetry) tolerance
3. Height gage to hold the dial indicator
4. Two precision parallels
5. Gage block stack that is equal to the actual length of datum A

STEP 1 Carefully rotate and slide the inspection equipment onto the surface plate.

STEP 2 Mount the part and the gage block stack between the two precision parallels, as shown in Figure 17.3. The gage block stack is being used to maintain the distance between the parallels for datum contact since the part is thin and would be unstable.

Figure 17.3 The part and gage block stack are mounted between two precision parallels (to establish datum centerplane A).

STEP 3 Measure the distance from the bottom parallel surface in contact with the part to the top of the tab, as shown in Figures 17.4 and 17.5.

STEP 4 Carefully rotate the part 180° and replace it between the parallels. Repeat step 3 to obtain the opposite distance.

STEP 5 Evaluate the differential measurements. Note that if the tab centerplane were perfectly symmetrical to centerplane A, both of the previous measurements

Figure 17.4 Zero is set on the surface of the bottom parallel.

Figure 17.5 The distance is measured from the surface of the bottom parallel to the opposite tab surface.

would be identical (the differential would be zero). To evaluate the measurement of symmetry in this case, find the difference between the two measurements and divide the difference by 2. The result of this is the direct amount that the centerplane of the tab is off from the datum centerplane. According to the tolerance on this particular part, the centerplane is allowed to be off by .002″ in either direction (.004″ total wide zone). If the differential does not exceed .002″ on this part, it is acceptable.

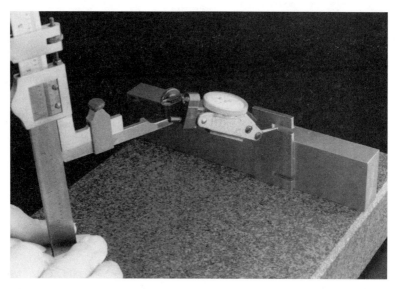

Figure 17.6 Setup for symmetry (position) inspection on the part locating from one side at a time.

An alternative method would be to simply stand the part up vertically on the surface plate and make two measurements while locating only one side of datum feature A. This method would have some error associated with it unless the surfaces of datum feature A were parallel within about 10% of the position tolerance. This setup is shown in Figure 17.6.

REVIEW QUESTIONS

1. Symmetry requires a _____ reference for measurement.
2. Symmetry tolerance is always applied using the _____ modifier.
3. Inspection of symmetry requires _____ measurements, not direct measurements.
4. Symmetry applies to the _____ line or _____ plane of the size features being controlled.
5. The tolerance zone for symmetry is _____ _____ planes that are the stated tolerance apart.
6. The symmetry symbol was deleted in the _____ revision of the standard, and then reinstated in the 1994 revision.
7. In order to control the location of symmetrical features using the MMC modifier, the option to the symmetry symbol is a _____ tolerance.
8. More often than not, symmetry applications are for _____ purposes as opposed to functional purposes.
9. The application of symmetry is to ensure that the controlled features are _____ in distance from the datum.
10. Symmetry cannot be inspected using a functional gage. *True* or *False?*

18

Introduction to Functional Gage Design

INTRODUCTION

Functional gages have been referred to as "precision worst-case mating parts," "receiver gages," or "function gages." The primary purpose of a functional gage is to gage the worst-case (virtual) boundary of the part while simulating the configuration of the mating part. Functional gages are applicable only to geometric tolerances that apply at maximum material condition (MMC). Tolerances that are modified by regardless of feature size (RFS) or least material condition (LMC) cannot be gaged with fixed gages.

FUNCTIONAL GAGE DESIGN PRINCIPLES

Some basic principles apply during the design of a functional gage:

1. Functional gages must be designed with features that will correctly establish all applicable *datums*. Gage planes that are flat, parallel, and/or perpendicular within gaging tolerances are used to contact part datum features and establish datum planes. Gage pins must contact part datum holes, and gage holes must contact part datum diameters.

2. Gaging members, such as gage pins, holes, bushings, and other plugs, must be designed to gage the *virtual boundary* of the part (which is generated by the size and geometric tolerance applied to the part features).
3. *Gaging tolerances* should not consume more that 10% of the tolerance being gaged. Refer to later sections of this chapter for more information.
4. *Wear allowance* must be established and designed into gage tolerances in order to ensure that the gage will last during production. Refer to later sections of this chapter for more information.
5. *Gage materials* should be selected based on part materials being gaged, use of the gage, required stability, and other factors.
6. The *configuration* of the gage for a given part should be a simulation of its mating part.
7. A functional gage *does not* consume the *intended clearance* between two mating parts. It gages only the virtual condition of the part.
8. A functional gage is not a "*go*" or "*no-go*" gage. It is one gage that either goes in/on the mating part, or does not. The only exception to this occurs when the part tolerance is specified using the *zero tolerance at MMC* method. When this method is used, the MMC size of the controlled feature and the virtual condition are the same boundary. Therefore, in this specific case, the gage member serves as the functional acceptance gage and the "go" size gage.
9. Functional gages do *not* gage the *size* of part size features. Size gaging must be accomplished with other gages.
10. Functional gages are designed specifically to *reject borderline-good* parts.

FUNCTIONAL GAGE ADVANTAGES AND DISADVANTAGES

There are a variety of factors to consider in determining whether to use a functional gage for inspection of the product. The paragraphs that follow cover some of the main advantages and disadvantages to using functional gages.

Advantages of Functional Gages

The following list identifies some of the most significant advantages of using functional gages to inspect geometric tolerances (if specified at MMC):

1. The product can be inspected quickly through the use of functional gages.
2. If the product fits the gage, it will fit every mating part (as long as the mating part is within specifications).
3. Functional gages represent the function of the part.
4. Functional gages are typically very easy to use during or after production.
5. Functional gages are always designed specifically so that they cannot possibly accept a defective part, but they can reject a borderline-good part. This fact will be discussed in more detail later in this chapter.

6. For very "tricky" tolerances—such as best-fit patterns, additional tolerances from size datums, virtual-condition datums at MMC, and others—functional gages solve many open-setup or coordinate measuring machine (CMM) inspection problems.

Disadvantages of Functional Gages

The following list identifies the main disadvantages of using functional gages:

1. They can reject borderline-good products due to wear allowance.
2. They are hardened gages that must be reworked or changed whenever associated drawing tolerances change.
3. There are costs associated with the design, fabrication, storage, and maintenance of functional gages. The largest cost is often in fabrication. Hardened steel gages made to very close tolerances are usually expensive.
4. Fixed gages do not give quantifiable results. If one needs variable measurements for certain procedures such as process control or first article inspections, functional gages will not provide data on variables. The only result of functional gaging is to accept or reject.

FUNCTIONAL GAGING COSTS

Numerous costs are associated with functional gaging:

1. The cost of designing and preparing design drawings for the gage.
2. Administrative costs, which include preparing purchase orders, receivers, and other documents associated with purchasing the gage.
3. The cost of gage materials and gage manufacturing. Gages can cost a considerable amount of money because of material costs and tight tolerance manufacturing methods.
4. Gage inspection can often be a problem because of the accuracy required when inspecting tight tolerance gages.
5. Gages must be certified on a specific certification frequency where the gage is inspected with respect to its design and changes are made to make it useful when rejected.
6. New gaging members (such as gage pins, bushings, or tabs) must be purchased when old members are past their "wear allowance." In other cases, gage members can be plated and ground back to original size. These operations can also be costly.
7. Storage and maintenance of gages is a relatively small expense, but deserves attention on the list of associated gaging costs.
8. If the production part design is changed, gages must often be reworked to meet the revised requirement, or new pins, etc. At times the gage is scrapped because it is no longer useful in the new part design.

BASIS FOR DECISIONS TO USE FUNCTIONAL GAGES

Other factors are also involved when making the decision to use a functional gage.

Fundamental Decisions
1. Part tolerances must be specified at MMC or functional gages cannot be used.
2. Attribute (accept or reject) results are all that is needed. Functional gages do not provide data on variables (measured values).

Technical Decisions
1. The tolerance of the part may be prohibitive to gaging. When the part itself is closely toleranced, gage designs become a problem because gages are produced to consume only 10% of the part tolerance.
2. Part materials may prohibit gaging because the materials are so abrasive that gage members wear out quickly unless they are hard enough (or plated) to withstand abrasive parts.

Cost-Related Decisions
1. The cost of gages are often too great and difficult to justify unless the product being gaged is long-term (mass production). One does not design and use functional gages if the entire order consists of only a few parts.
2. Refer to all the cost areas previously covered for more examples in which costs prohibit the purchase of a functional gage.

MATERIALS USED FOR FUNCTIONAL GAGES

A variety of materials can be used for manufacturing functional gages. Some of the most important material-related facts to consider are *hardness, toughness, wear resistance, machinability, tensile strength, shear strength, and brittleness.* Many fixed gage applications use carbon steel. Other applications may require better metals such as tool steels for wear resistance and stability reasons (such as very tight part tolerances or abrasive part surfaces).

Hardness is the ability of a material to resist penetration or indentation. The harder the material, the greater the tensile strength.

Toughness is the ability of a material to absorb sudden applied loads or repeated shocks without becoming permanently deformed. For gage materials, on a hardness scale, toughness is roughly between the Rockwell "C" scale hardness of RC 44 to RC 48. Above RC 48, brittleness tends to replace toughness.

Wear resistance is the ability of a material to resist abrasion. Hardness is a main factor in wear resistance.

Machinability is a term used to describe how easy the material is to machine.

Brittleness is basically the opposite of toughness. Brittle materials tend to fracture when sudden loads are applied.

Tensile strength is the ability of a material to resist being pulled apart. Tensile strength is the primary test used to determine the strength of materials. As the hardness of a material increases, tensile strength tends to increase proportionally.

Shear strength is the ability of a material to resist forces applied in opposite and parallel directions. Shear strength tends to be approximately 60% of the tensile strength.

Carbon Steels

Carbon steel, whether it has low-, medium-, or high-carbon content, is the primary material used in most jig and fixture designs because of its machinability, low cost, availability, and versatility. Low-carbon steel contains from .05% to .30% carbon. It is used mainly for structural parts of fixtures and gages such as base plates, gage bodies, or supports. Medium-carbon steel contains from .30% to .50% carbon.

Carbon steel is used in areas of the gage or fixture that require great strength such as clamps, studs, or nuts. Medium-carbon steel, however, is more difficult to machine, case-harden, or weld. High-carbon steel contains from .50% to 2.0% carbon. High-carbon steel is used in areas of gages and fixtures that are subject to the most wear and tear, such as drill bushings, locators, wear pads, supports, and gage pins.

Tool Steels

Gage blocks, for example, are made of tool steel material that is hardened tool steel (Rockwell C 62 minimum). The drawbacks to using a tool steel is its susceptibility to scratches, nicks, and corrosion in certain use environments.

Tool steels come in various grades. Grade A tool steel is often used for functional gage designs.

W—Water-hardened

D—High-carbon, High-chromium

A—Medium alloy, Air-hardened

O—Oil-hardened

T—Tungsten base

M—Molybdenum base

F—Carbon-tungsten base

Chrome-Plated Gaging Members

At times some applications call for chrome (chromium) plating on certain gaging members (such as gage pins or rings that make direct contact with the part during inspection). Chrome plating resists the problem discussed previously with tool steels, but it is also an expensive venture. Chrome plating applies to surfaces and members that come in contact with the part. It reduces friction during gaging and therefore increases the wear resistance of the gage member. Chrome plating can also be used to repair or restore worn gage pins and rings. The pin (or ring) that has worn away can be plated and machined back to its design size.

Carbide Steels

Carbide steels (whether chromium carbide or tungsten carbide) are the hardest. They are almost immune to scratches or nicks and have a very high resistance to corrosion. The hardness of carbide steels can allow them to outlast other steel gages by about 15 to 20 times, but the cost of carbide steels can be far higher than the cost of carbon steels. Tungsten carbide or chromium carbide is commonly used for specific applications where the benefits outweigh the cost. Both carbides have the following characteristics in common:

1. They provide greater stability and wear resistance than common tool steels.
2. Both are sensitive to chipping.

Ceramics

Ceramic materials are gaining popularity in tool and gage design. Ceramics provide a higher degree of wear resistance, but are also prone to brittleness (sensitive to chipping).

SURFACE ROUGHNESS (TEXTURE)

Surface roughness (or texture) of gaging members plays a significant role in the wear of gaging members and interface surfaces. Gage manufacturers use special lapping processes to produce a smooth finish. These lapping processes also improve the geometrical error of gaging members (such as circularity, taper, straightness, or flatness).

GAGE MAKERS' TOLERANCE AND WEAR ALLOWANCE

In this chapter, certain gage designs for different requirements will be introduced. It must be understood that whenever gaging surfaces are mentioned, they are intended for locating part datum features. These datum-locating surfaces must be mutually parallel, perpendicular, and angular to each other within gage tolerances. It is also understood that whenever gage pins or bushings are used to gage controlled features of the part, they must also have gage makers' tolerance and wear allowance. These tolerances and wear allowances cause gage pins, for example, to be equal to or a bit larger than the virtual condition of the hole in the part. For this reason, functional gages can always reject a part when the part is borderline-acceptable. Gage tolerances depend on part tolerances.

The tolerance of the part plays a major role in the decision to use functional gages. In order to prevent significant gaging errors, functional gages should not consume more than 10% of the part tolerance being gaged. If possible, less than 10% is better. It is also important to note that the 10% consumed by the gage should be the accumulation of all associated gage tolerances; therefore, the stackup of gage tolerances needs to be considered. Some part tolerances are so tight that gaging manufacturers cannot produce gages to the 10% rule. Gaging error, in inspection, can be explained by two primary errors: alpha risk or beta risk. The *alpha risk* is the probability of rejection of good products. The *beta risk* is more serious; it is the probability of acceptance of defective product.

Fixed gages are always "*plus against the part*," meaning that part of the gage is always rubbing against features of the part. Gages are designed specifically for alpha risk. They are intended to reject good products that are right at the borderline of their tolerance limits. All gages must have gage makers' tolerance to manufacture gage and wear allowance for using the gage. Due to fact that the tolerance for both the gage maker and the wear allowance are additive, gages must reject borderline-good products. The beta risk is never used in gage design. Acceptance of defective product is not a viable option. When using functional gages, one should always remember that these gages can make acceptance decisions, but they cannot make the final rejection decision. If a gage rejects a part, the part should be inspected in other ways that will confirm the rejection or accept the part at the borderline.

Establishing Gage Wear Allowance

With all gages, wear allowance is necessary. Functional gages are often used extensively in manufacturing, and each use contributes to the wear of gaging surfaces and members (such as gage pins, bushings, etc.). During the design of the gages, the 10% rule is applied and gage tolerances must be decided. Inclusive in gage tolerancing is the allowance for wear. Wear allowance adds size to gage pins and reduces the size of the gage holes. Wear allowance specification is determined by considering the gage material selected, the material of the part, the quantities of parts that might be gaged, and the type of gaging operation. Gage makers should be consulted regarding gage tolerances and wear allowances for specific applications.

GEOMETRIC TOLERANCES THAT CANNOT BE "GAGED"

The only geometric tolerances that can be functionally gaged are those which apply at MMC. Geometric tolerances that are applied at RFS or LMC are not conducive to functional gaging. Also, any geometric tolerance that applies to a surface, or to a line element of a surface, cannot be functionally gaged. The following is a list of geometric tolerances for which functional gaging is prohibited (although some form of variable gage might apply that includes dial indicators or probes):

1. Flatness
2. Straightness of surface elements

3. Straightness of an axis (or centerplane) — RFS
4. Angularity of a surface
5. Perpendicularity of a surface
6. Parallelism of a surface
7. Circularity (roundness)
8. Cylindricity
9. Profile of a line or surface
10. Runout (circular and total)
11. Concentricity
12. Symmetry
13. Size features controlled with any geometric tolerance on an RFS or LMC basis

The Effect of "Tight" Tolerances on Gage Designs

Tight part tolerances can cause extremely tight gage tolerances. For example, a single hole has a position tolerance on the part drawing of a .001″ cylindrical zone at MMC with respect to specified datums. The total tolerance for position is .001″ cylindrical zone. During the start of gage design, the gage designer calculates 10% of the position tolerance (10% of .001″ = .0001″). This .0001″ is all the tolerance allowed for the functional gage.

Tight part tolerances have the following general effects on gage design:

1. Increased cost for gage fabrication due to very tight gage tolerances.
2. Some parts cannot be gaged because there is nothing left for gage makers' tolerance or wear allowance.
3. Gage inspection, once the gage is completed, becomes another problem because the inspection of the gage should not consume more than 10% of gage tolerance. Therefore, 10% of .0001″ equals .00001″. This could create the need for very expensive inspection equipment (such as one of the best coordinate measuring machines).

PART DATUMS AND FUNCTIONAL GAGE INTERFACES

Functional gages must be designed to make proper contact on all part datum features before inserting gage members in the controlled features of the part. For example, the gage body must properly locate the datum features of the part before inserting the gage pin in the hole to check its position. Plane datum features are part surfaces where the high points must be contacted, as shown in Figure 18.1. The primary datum must be established by contacting the three or more highest points on the primary datum feature.

The secondary datum is then contacted at the two (or more) highest points. Lastly, if a tertiary datum exists, it must be contacted at one or more of the highest points on that surface. Gage surfaces must be designed such that the surfaces have a large-enough area to contact the part datum features completely. In most cases, gaging surfaces that establish datums are expected to be closely lapped (flat) surfaces.

312   Chapter 18   Introduction to Functional Gage Design

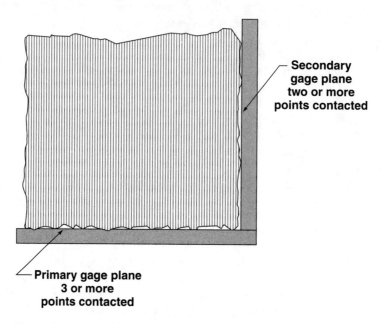

Figure 18.1 Gages must have planes to contact plane feature datums.

FUNCTIONAL GAGE INTERFACES FOR SIZE FEATURE DATUMS

In cases where part datums are size datums (such as a hole, slot, pin, or boss), a representative gage member (within gage tolerances) must be designed to contact size feature datums. For example, a datum hole specified at MMC on the part requires that a gage pin be designed that is at MMC size within gaging tolerances (plus wear allowance). In this manner, any additional tolerance is accounted for due to the clearance between the gage member and the actual size of the datum feature.

Now look at Figure 18.2. Notice that datum A is a hole with a size tolerance of between 2.005″ and 2.007″ and that it is the primary datum for the four-hole pattern. The position tolerance for the four holes specifies datum A at MMC. The MMC size for datum A is the smallest hole (2.005″). The gage design shows that a pin must be at the MMC size of datum A (plus wear allowance, of course).

SIZE FEATURE DATUM–VIRTUAL CONDITION RULE

Size feature datums are datums of size that are called out at MMC, but at times, a virtual condition exists because the datum feature is controlled by another geometric tolerance to other datums. In these cases, the virtual condition of the size datum feature is the basis for the relationship, not the MMC size. Refer to the ***datum–virtual condition rule***. Consider Figure 18.3; this figure is the same as the previous part (see Figure 18.2) except that datum A now has a perpendicularity tolerance of .003″ to datum B. The

Size Feature Datum–Virtual Condition Rule

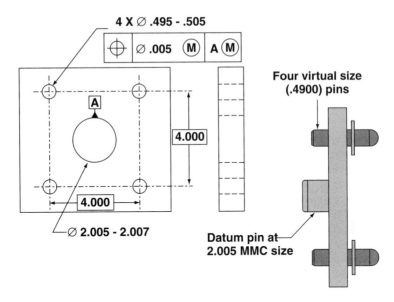

Figure 18.2 A part with a size datum called out at MMC.

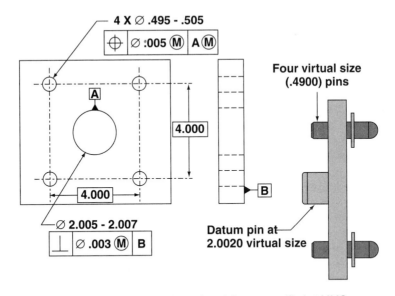

Figure 18.3 Functional gage for a virtual datum specified at MMC.

MMC of datum *A* is still 2.005″, but it is allowed to be out of perpendicularity up to .003″ at MMC size.

The virtual condition of datum *A* is now MMC (2.005″) minus the perpendicularity tolerance (.003″), which equals 2.002″. Whenever a datum of size has a worst-case (virtual) condition, the virtual condition is the basis for the relationship, not the MMC size. In this case, a virtual pin of 2.002″ is used in the gage to establish datum A for the position tolerance.

GAGE PINS

A variety of types of locator or gaging pins may be used in production fixtures, inspection fixtures, or functional gages, as shown in Figure 18.4. The rounded or bullet pin configuration, for example, is often used in gage designs where fixed pins are in the gage body to mount datums of size. This is so because the pin configuration makes it much easier to put the part on the gage and to remove the part. All gage pins should have an end configuration that makes it easier for the pin to start in the hole.

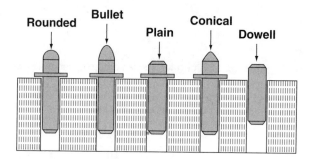

Figure 18.4 Various locator and gage pin configurations.

At times, functional gages must contact datum targets (such as target points, lines, or areas). In these cases, gage locator pins are necessary in the functional gage. The rounded or bullet pin design is used to establish datum target points of contact, the plain pin design can be used to establish datum target areas, and the outer edge of a gage pin can be used to establish datum target lines of contact.

Pins can also be used in functional gage designs (or simple inspection fixtures) for support, location, alignment, stops, or error proofing.

FIXED PINS VERSUS SLIDING PINS

Sliding-pin gages (those which have removable pins) provide certain benefits over fixed-pin gages:

1. Proper contact can be made on datums; then the pin(s) are inserted.
2. Features out of tolerance can be evaluated to determine which one is out and an idea of the direction in which it is out.

Sliding-pin designs can use a hole or a gage bushing. Gage bushings are often recommended because they can be replaced more economically than a gage body. It is recommended that part tolerances with projected tolerance zones should be used on the bushing that holds the sliding pin. Fixed-pin gages are often less expensive, but problems arise when attempting to contact datums or to evaluate which features are the cause of rejection.

FUNCTIONAL GAGE DESIGN EXAMPLES

The following are examples of various functional gage designs for geometric tolerances specified at MMC. It is important to keep the following general criteria in mind during gage design:

1. Gages must properly contact part datums at all times.
2. A functional gage is similar in configuration to the mating part, and yet it is designed to check the virtual conditions of the intended part.
3. The 10% rule applies to selecting gage tolerances. The intention is that 10% of part tolerance is consumed by the stackup of gage tolerances.
4. Wear allowance must be factored to ensure that the gage can withstand repeated use on production parts and that it does not consume too much of the part tolerance.
5. Functional gages do not check the size of controlled features. They must always be accompanied by size gages or measurements.

Perpendicularity of a Hole at MMC

The drawing in Figure 18.5 shows a hole that must be perpendicular to primary datum A within .002″ when it is produced at MMC size (.480″ diameter). The functional gage

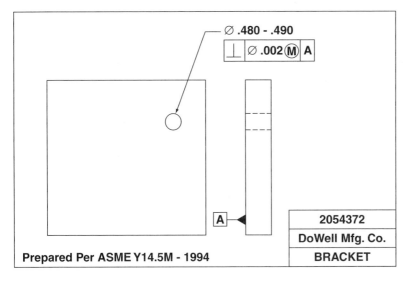

Figure 18.5 A perpendicularity tolerance for a hole specified at MMC.

is designed to have a flat gage plane that will contact all of datum A, and then a virtual size gage pin must pass through the hole. The gage pin design should use a sliding pin so that it can be retracted until all of datum A has been contacted. Then the gage pin must pass through the hole. The size of the gage pin must be .4780″ (allowing for gage makers' tolerance and wear allowance).

Figure 18.6 shows a sketch of the basic design of the gage.

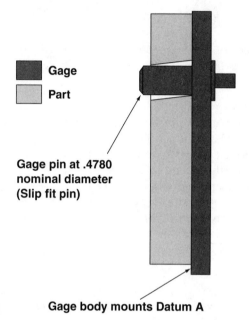

Figure 18.6 Functional gage for perpendicularity at MMC.

Straightness of an Axis at MMC

The drawing in Figure 18.7 specifies that the axis of the shaft must be straight within a cylindrical tolerance zone of .002″ when the shaft is produced at MMC size (.375″ diameter). The virtual condition of the shaft is .375″ (MMC) plus the straightness tolerance (.002″), which equals .377″. The functional gage, as shown in Figure 18.8, is a ring gage that is longer than the shaft and has a bore diameter of .3770″ within gaging tolerances. Due to wear allowance, it might even be smaller than .3770″.

Position of a Hole to Three Datum Planes

Another example, shown in Figure 18.9, is a position tolerance that has been applied to the location of a hole when the hole is at MMC size. In this case, the gage design must include three gage surfaces to contact datums A, B, and C. Then, a slip-fit pin of .4780″ diameter must pass through the hole (as shown in Figure 18.10). Notice, in this

Functional Gage Design Examples 317

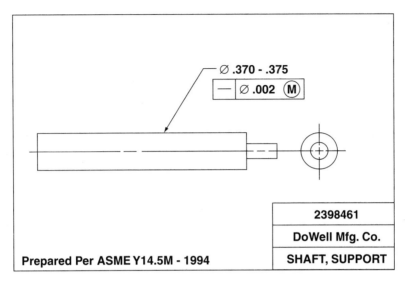

Figure 18.7 Drawing for straightness of an axis at MMC.

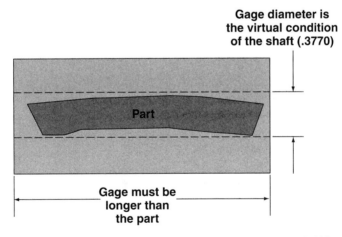

Figure 18.8 Functional gage for straightness of an axis at MMC.

Chapter 18   Introduction to Functional Gage Design

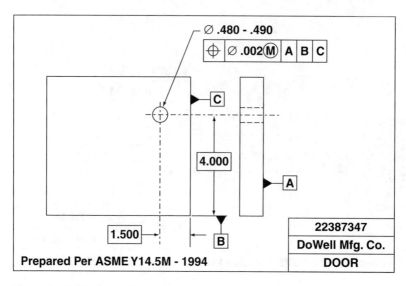

Figure 18.9   Position of a hole to three datum planes.

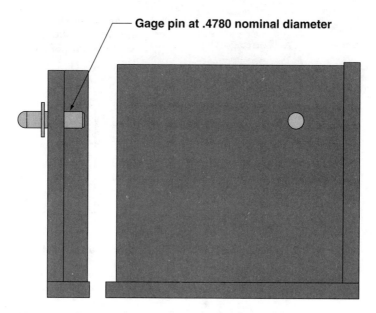

Figure 18.10   Functional gage for hole position at MMC.

case, that a slip-fit pin is appropriate because a fixed-pin design may prevent the part from contacting the datum-related surfaces of the gage.

Coaxial Position Tolerance

Figure 18.11 shows another type of position tolerance that is used only to control the coaxiality of the three holes to one another. The requirement is an "intrinsic" datum, which means that the three holes are related only to one another. The MMC size of the three holes is .370″ diameter. The virtual-condition boundary that must be inspected by the functional gage is MMC (.370″) minus the position tolerance (.003″), which equals .367″. In this case, the functional gage is a pin that is .3670″ diameter (allowing for gage makers' tolerance and wear allowance). The gage is shown in Figure 18.12.

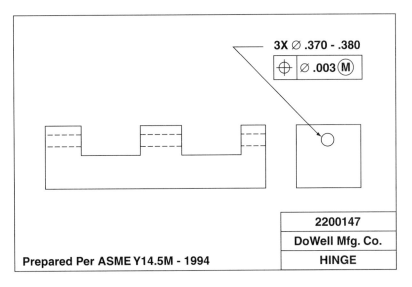

Figure 18.11 Coaxial position tolerance at MMC.

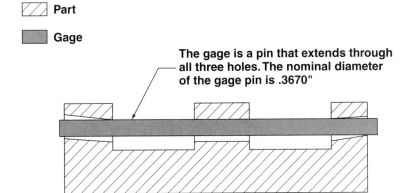

Figure 18.12 Functional gage for coaxial position tolerance at MMC.

Coaxial Position Tolerance and Pattern Location

Figure 18.13 shows a different requirement for the previous three-hole pattern. In this case, the part interfaces with the mating part on datums *A* and *B*, and so the three holes in the pattern are related to datums *A* and *B* and to one another.

The functional gage, shown in Figure 18.14, must have two gaging surfaces that are flat and perpendicular to each other within close tolerances. It then may have two bushings at each end to support the gage pin (.3670″ diameter) while it is passed through the three holes.

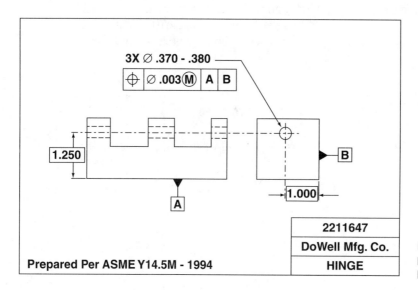

Figure 18.13 Coaxial position tolerance and pattern location.

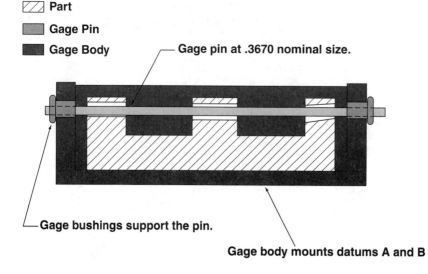

Figure 18.14 Functional gage for coaxial position tolerance and pattern location.

Position Tolerance — Intrinsic Datums

As covered in Chapter 16, position tolerances of hole patterns that have intrinsic datums (the holes to one another) allow for best-fit acceptance during inspection. "Best fit" is the ability to rotate or shift inspection measurements to see whether the holes are within their tolerance zones. The drawing in Figure 18.15 is a position tolerance where best fit is allowed. Normal inspection methods using height gages, surface plates, or coordinate measuring machines cannot make the proper acceptance decision on this part unless they are accompanied by graphical inspection analysis. Functional gages can make the best-fit decision because the four holes will share their clearance with one another and the four gage pins will accept the part.

The drawing shows that the holes must be in position to one another and perpendicular to the primary datum. The virtual condition of the four holes is MMC (.495" diameter) minus the position tolerance (.005" cylindrical zone), which equals .4900".

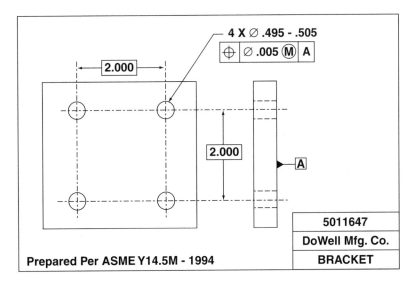

Figure 18.15 Position tolerance of a four-hole pattern (intrinsic datum).

The gage design has one surface for mounting the primary datum and four slip-fit pins at .4900" diameter (allowing for gage makers' tolerance and wear allowance). At this point, it is also important to note that the locations of the bushings for the gage pins have to share the same basic dimensions that locate the zones for the four holes in the part. This example is a good time to cover gage tolerances and gage tolerance stackup.

If the 10% rule is followed, the tolerance of the gage will be 10% of .005" position tolerance (or .0005"). The .0005" is all of the tolerance of the gage, which includes the location tolerance of the four pins, tolerances accumulated if gage bushings are used, and finally, the size tolerance for each gage pin.

322 Chapter 18 Introduction to Functional Gage Design

In gage design, it is often true that location tolerances are more difficult to hold for a gage manufacturer than the size tolerances of the gage members. In this case, it could be that between .00035" and .0004" of the tolerance will be used for gage pin or bushing locations (to each other) and the remaining .0001" to .00015" tolerance would be used for gage pin size. Then the amount of wear allowance must still be considered (which will add to the size of the gage pin). An example of the functional gage for this part is shown in Figure 18.16.

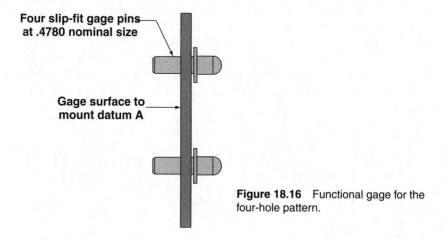

Figure 18.16 Functional gage for the four-hole pattern.

Position Tolerance and Pattern Location

Figure 18.17 shows a part drawing with a position tolerance that has been called "one single-segment" position (in the 1994 standard). The four imaginary tolerance zones are .005" if the hole is produced at MMC size. These zones are not only located to one another, but the pattern of zones is located to datums A, B, and C. The gage design, as sketched in Figure 18.18, would normally be a slip-fit pin design that shares the same basic dimensions for the part.

Gage tolerance stackup is compounded by the location of the gage pin bushings from each other, the location of all bushings from datums A, B, and C, the size tolerance of the gage pins, and the wear allowance that is determined. The use of the gage is to retract the four pins, mount the part datums on the gage, and then attempt to insert the four pins in their respective holes.

ALTERNATIVES TO FUNCTIONAL GAGING

Whenever the decision is made not to use a functional gage, a variety of options for inspection of the geometric tolerance are available:

1. Open-setup inspection using surface plates and accessories
2. A coordinate measuring machine (CMM) with geometric tolerancing software

Alternatives to Functional Gaging   **323**

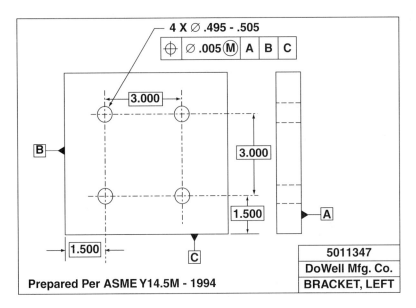

Figure 18.17   One single-segment position tolerance.

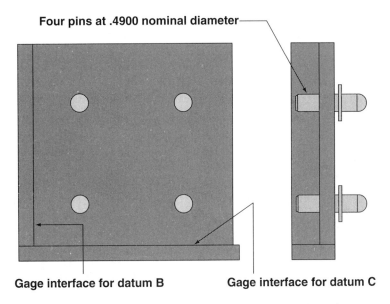

Figure 18.18   Functional gage for one single-segment position tolerance.

324 Chapter 18 Introduction to Functional Gage Design

3. Variable gaging fixtures (such as fixtures to locate the part that have probes or dial indicators to make direct measurements)
4. Other measuring machines such as profile analyzers, circularity measuring machines, and side-view comparators.

USING FUNCTIONAL GAGES

The purpose of a functional gage is to gage the virtual (worst-case) condition of the part. Functional gages are not size gages. Even when functional gages accept the part, one must measure the size of the feature separately (using "go" and "no-go" gages or measuring instruments). The exception to this occurs when a functional gage has been designed to gage a part that has *zero at MMC* tolerance. In this case, the gaging member of the gage is at the virtual condition and the virtual condition is the same as the MMC condition of size. Therefore, when zero at MMC tolerance is applied, the gaging member (pin or ring) is also the "go" size gage. (See Figure 18.19.)

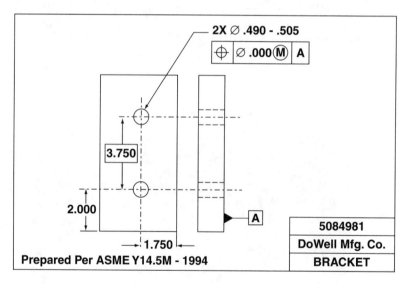

Figure 18.19 A two-hole pattern with zero at MMC position tolerance applied.

The drawing in Figure 18.19 shows a two-hole pattern that has been positioned using zero at MMC tolerance method. Since the MMC of the two holes is .490″ and the virtual condition is the same (MMC .490″ minus the position tolerance .000″ = .490″), the pins in the functional gage (.4900″) not only serve as the functional gage pins, but also the "go" size gage. The functional gage for the part, shown in Figure 18.20, has an interface surface to mount primary datum *A* and two gage pins that are .4900″ in size and located 3.7500″ from each other within applicable gaging tolerance.

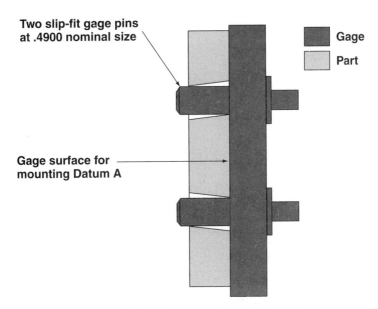

Figure 18.20 Functional gage for the two-hole pattern position tolerance.

Another aspect of functional gaging is the amount of force to apply to gage pins (or other gaging members). A functional gage pin should go through the hole under its own weight. No specific force should be applied. If the user is tempted to use force on the pin, the part is probably one of those that are acceptable at the borderline of acceptance. Remember, functional gages are supposed to reject borderline-good products. Instead of force on the pins, or rejecting the part arbitrarily, another method should be used to make the final decision.

EXAMPLES OF THREE ACTUAL GAGE DESIGNS

Part 1: Position Tolerance—MMC Size Datum

The first part is an example of a position tolerance on a four-hole pattern at MMC with respect to a size datum specified at MMC (see Figure 18.21). Notice that, for this part, the primary datum is the inside diameter and the secondary datum is the face. Due to the fact that the face is a secondary datum, the gage can be designed with fixed pins. Figure 18.22 shows two views of the actual part.

The gage design drawing for the position tolerance is shown in Figure 18.23. Notice that the gage design drawing uses the old coordinate system. Although geometric tolerancing could also be used on gage design drawings, it is typical in the industry to use the coordinate system. If one calculates 10% of the position tolerance on the part,

Chapter 18 Introduction to Functional Gage Design

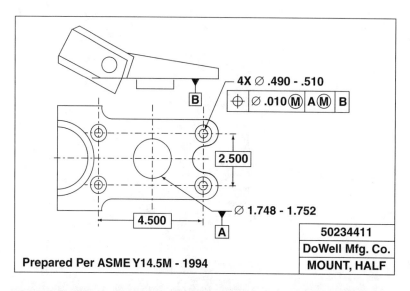

Figure 18.21 The drawing for part 1 position tolerance.

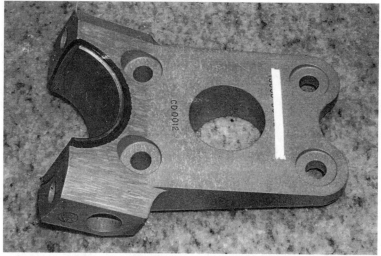

(a)

(b)

Figure 18.22 Two views (a and b) of part 1 position tolerance. (Courtesy of Barry Controls Aerospace, Burbank, California.)

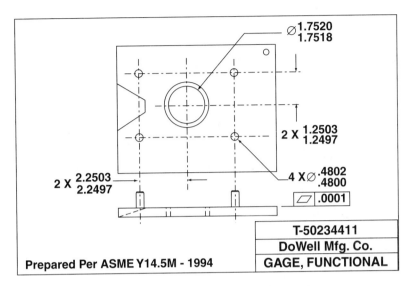

Figure 18.23 The drawing for the functional gage for part 1.

the gage tolerances should consume only .001″ (which is 10% of the .010″ position tolerance). Further, if one calculates the stackup of the gage dimensions, it will show that

1. Plus or minus .0003″ has been assigned to the location of the gage pins (Note that this tolerance has a Hypotenuse of .00042″, which is the worst case.)
2. The pin size tolerance is plus or minus .0001″
3. The datum-locating hole is plus or minus .0001″

The stackup of the gage tolerances consumes less than .001″, and the gage pins have been assigned .0002″ wear allowance. The actual functional gage is shown in Figure 18.24.

When using this gage to inspect the part, the observer merely applies the part to the gage carefully and ensures that, without force,

1. The four holes go over the gage pins
2. The datum diameter goes into the datum contacting hole
3. The primary datum interfaces with the face of the gage

Figure 18.25 shows that the part has been accepted by the functional gage.

Part 2: Position Tolerance—Virtual Condition Datum

The next example part (shown in Figure 18.26) is also a part with a four-hole pattern position tolerance. In this case, however, the size datum has a *virtual condition* due to the perpendicularity tolerance that controls datum *B* to datum *A*. Referring to the *datum–virtual condition rule*, the virtual condition of datum *B* (which is 2.990″) is the basis for locating the pattern. In this case, the diameter in the gage that contacts datum

328 Chapter 18 Introduction to Functional Gage Design

Figure 18.24 The functional gage for part 1. (Courtesy of Barry Controls Aerospace, Burbank, California.)

(a)

(b)

Figure 18.25 Part 1 is shown on the gage in views (a) and (b). (Courtesy of Barry Controls Aerospace, Burbank, California.)

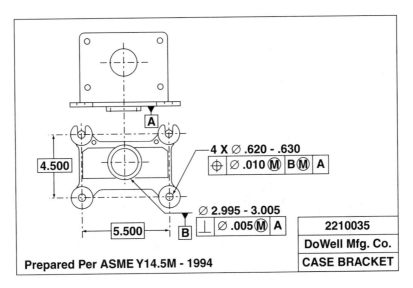

Figure 18.26 The drawing for part 2.

B must be based on 2.990″, not the MMC size of 2.995″. It is also important to note that two of the holes in the pattern have been relieved so that they can be used as open slots. The part drawing still treats them as holes in the pattern.

The actual part is shown in two views in Figure 18.27. The bottom view shows the four holes, the datum diameter, and the face involved with the position tolerance.

The gage design drawing is shown in Figure 18.28. This design has also been toleranced based on the 10% rule. The virtual condition on datum B has been accounted for, which will allow any additional shift in the four-hole pattern based on the difference between the virtual condition (3.010) and the actual virtual size of datum B when it is produced. The actual functional gage for the part is shown in Figure 18.29, and the part on the gage is shown in Figure 18.30.

Part 3: Position of a Two-Hole Pattern to a Primary Datum Face

The drawing for part 3 is shown in Figure 18.31. This part has a two-hole pattern that must be within position tolerance of .005″ at MMC with respect to primary datum face A and secondary datum diameter B at MMC. The position tolerance controls the location of the two-hole pattern from the datums and the holes to each other. Bonus position tolerance is allowed on either hole if it departs from MMC size, and an additional pattern shift from datum B is allowed if datum B is produced smaller than MMC size.

The functional gage drawing for part 3 is shown in Figure 18.32. The part position tolerance is .005″; therefore the gage should not consume more than 10% (.0005″). The gage has a plane to mount datum A, a hole to locate datum B, and two sliding gage pins at the virtual condition of the holes. The fit between the gage pin holes and the pins is a running (slip) fit.

330 Chapter 18 Introduction to Functional Gage Design

(a)

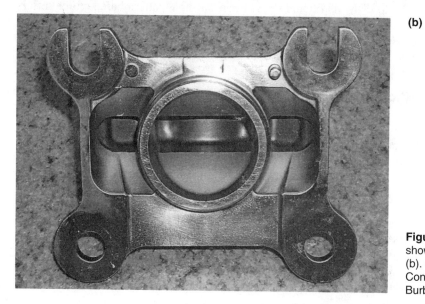

(b)

Figure 18.27 Part 2 is shown in two views (a) and (b). (Courtesy of Barry Controls Aerospace, Burbank, California.)

Examples of Three Actual Gage Designs 331

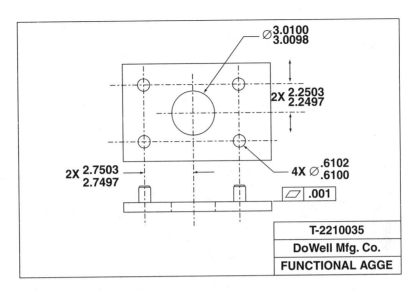

Figure 18.28 The drawing for the functional gage for part 2.

Figure 18.29 The functional gage for part 2. (Courtesy of Barry Controls Aerospace, Burbank, California.)

Chapter 18 Introduction to Functional Gage Design

Figure 18.30 Part 2 is shown on the functional gage. (Courtesy of Barry Controls Aerospace, Burbank, California.)

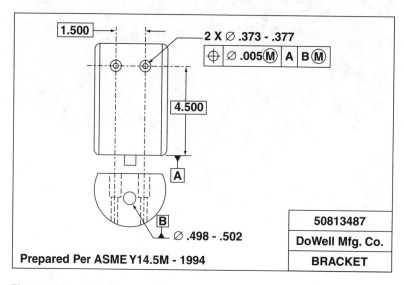

Figure 18.31 The drawing requirement for part 3.

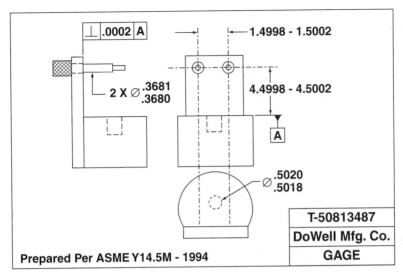

Figure 18.32 The drawing for the functional gage for part 3.

Figure 18.33 presents a picture of the functional gage. The part is not shown, but the use of the gage is a straightforward process. The two sliding gage pins are first retracted. The part is then mounted on the gage on datum *A* against the plane with datum *B* fully inserted into the gage hole. Once the part is mounted on the gage, the two pins are inserted into the holes without force.

COORDINATE MEASURING MACHINES (CMMS)

CMMs are very versatile. A wide variety of coordinate measurements can be made in at least three axes (*X, Y,* and *Z*). CMMs can simulate a functional gage if

1. Fixturing and gage pins are used
2. Graphical analysis is performed on the results

Benefits of Using CMMs

1. They can be used as a system for functional gaging. With the appropriate geometric tolerancing software, they can be used to simulate functional gaging on any part regardless of the tolerance modifier (RFS, LMC, or MMC).
2. They can be operated manually, computer-assisted, or computer-controlled.
3. They can reduce inspection fixtures and maintenance costs.
4. They reduce inspection setup time by means of automatic alignment.
5. They provide permanent records of inspection.
6. They can be set to operate in either English or metric units of measurement.

334 Chapter 18 Introduction to Functional Gage Design

Figure 18.33 The functional gage for part 3: (a) sliding pins retracted and (b) full insertion distance for the pins. (Courtesy of Barry Controls Aerospace, Burbank, California.)

Drawbacks of Using CMMs

1. The initial cost can be a major factor. CMMs, depending on the model purchased, can be very expensive. However, when used properly, they can pay for themselves in a very short time.
2. There can be environmental concerns such as measurement error if the CMM is not resting on a vibration-free foundation or if a temperature-controlled room is required.
3. CMMs are not portable. Product must be brought to the CMM. It cannot be taken to the process.
4. CMMs require a trained operator.

SUMMARY

Functional gages are used throughout the world by many manufacturing companies. Their ease of use, especially on the production line, makes them especially popular in mass production. One must be knowledgeable about the use, design, and drawbacks of functional gages in order to make the correct decision. Care must be taken to make certain that the gage design does not consume so much of the part tolerance that the gage rejects good products that are not at the borderline of acceptance.

REVIEW QUESTIONS

1. Functional gages can be designed for geometric tolerances that specify the MMC modifier. *True* or *False?*
2. Functional gages should consume only _____ percent of part tolerance.
3. Functional gages are designed to reject borderline good products. *True* or *False?*
4. Which of the following geometric tolerances cannot be gaged with a functional gage?
   a. Perpendicularity of a hole at MMC
   b. Position of a slot at MMC
   c. Perpendicularity of a surface
   d. Straightness of an axis at MMC
5. The worst-case condition (boundary) that functional gages are designed to check is called the _____ condition.
6. In cases where functional gages do not accept the part, which of the following decisions should be made?
   a. Reject the part.
   b. Apply more pressure to the gage pin.
   c. Inspect the part using another method or piece of equipment.
   d. Accept the part anyway.
7. What is the nominal size of a gage pin for the position of a diameter hole of between .495" and .505" with a position tolerance of .020" at MMC (before considering the gage makers' tolerance and wear allowance)?
8. Functional gage designs must be configured such that all datums are contacted properly. *True* or *False?*
9. When a functional gage is designed to gage holes in a part that has been positioned using the zero at MMC method, the functional gage pins also check the MMC size limit. *True* or *False?*
10. Which of the following factors are directly involved with establishing wear allowance for functional gages?
    a. The material from which the part is made
    b. The material from which the gage is made
    c. The volume of products that will be gaged
    d. All of the above

Appendices

APPENDIX A Conversion of Actual Coordinate Measurements to Position Location Diameter
(inches)

Y											
.020	.0400	.0402	.0404	.0408	.0412	.0418	.0424	.0431	.0439	.0447	.0456
.019	.0380	.0382	.0385	.0388	.0393	.0398	.0405	.0412	.0420	.0429	.0439
.018	.0360	.0362	.0365	.0369	.0374	.0379	.0386	.0394	.0402	.0412	.0422
.017	.0340	.0342	.0345	.0349	.0354	.0360	.0368	.0376	.0385	.0394	.0405
.016	.0321	.0322	.0325	.0330	.0335	.0342	.0349	.0358	.0367	.0377	.0388
.015	.0301	.0303	.0306	.0310	.0316	.0323	.0331	.0340	.0350	.0360	.0372
.014	.0281	.0283	.0286	.0291	.0297	.0305	.0313	.0322	.0333	.0344	.0356
.013	.0261	.0263	.0267	.0272	.0278	.0286	.0295	.0305	.0316	.0328	.0340
.012	.0241	.0243	.0247	.0253	.0260	.0268	.0278	.0288	.0300	.0312	.0325
.011	.0221	.0224	.0228	.0234	.0242	.0250	.0261	.0272	.0284	.0297	.0311
.010	.0201	.0204	.0209	.0215	.0224	.0233	.0244	.0256	.0269	.0283	.0297
.009	.0181	.0184	.0190	.0197	.0206	.0216	.0228	.0241	.0254	.0269	.0284
.008	.0161	.0165	.0171	.0179	.0189	.0200	.0213	.0226	.0241	.0256	.0272
.007	.0141	.0146	.0152	.0161	.0172	.0184	.0198	.0213	.0228	.0244	.0261
.006	.0122	.0124	.0134	.0144	.0156	.0170	.0184	.0200	.0216	.0233	.0250
.005	.0102	.0108	.0117	.0128	.0141	.0156	.0172	.0189	.0206	.0224	.0242
.004	.0082	.0089	.0100	.0113	.0128	.0144	.0161	.0179	.0197	.0215	.0234
.003	.0063	.0072	.0085	.0100	.0117	.0134	.0152	.0171	.0190	.0209	.0228
.002	.0045	.0056	.0072	.0089	.0108	.0126	.0146	.0165	.0184	.0204	.0224
.001	.0028	.0045	.0063	.0082	.0102	.0122	.0141	.0161	.0181	.0201	.0221
X ▸	.001	.002	.003	.004	.005	.006	.007	.008	.009	.010	.011

Example: An X coordinate difference of .005" with a Y coordinae difference of .009" equals a diametral zone of .0206". For those values not found in this table, use the following equation:

$$\text{Diametral zone} = 2\sqrt{\Delta_X^2 + \Delta_Y^2}$$

Note: This table can be expanded to larger values by moving the decimal point the same number of places in both coordinates and the answer.

APPENDIX B Conversion from Diametral Tolerance Zone to Plus and Minus Coordinate Tolerances

Diametral Zone (")	Plus and Minus Tolerance (")	Diametral Zone (")	Plus and Minus Tolerance (")
.001	.00035	.016	.00566
.002	.00071	.017	.00601
.003	.00106	.018	.00636
.004	.00141	.019	.00671
.005	.00177	.020	.00707
.006	.00212	.021	.00742
.007	.00247	.022	.00778
.008	.00283	.023	.00813
.009	.00318	.024	.00848
.010	.00353	.025	.00884
.011	.00389	.026	.00919
.012	.00424	.027	.00955
.013	.00459	.028	.00990
.014	.00495	.029	.01025
.015	.0053	.030	.01061

Example: A position tolerance diameter of .010" would be plus or minus .00353" on each coordinate.

Note: To expand this table, move the decimal point the same amount of places in the diameter and in the plus or minus tolerance value found.

The following equation can be used for values not shown:

$$\text{Plus/minus tolerance} = \frac{.7071" \times \text{diametral zone}}{2}$$

APPENDIX C Bolt Circle Chart

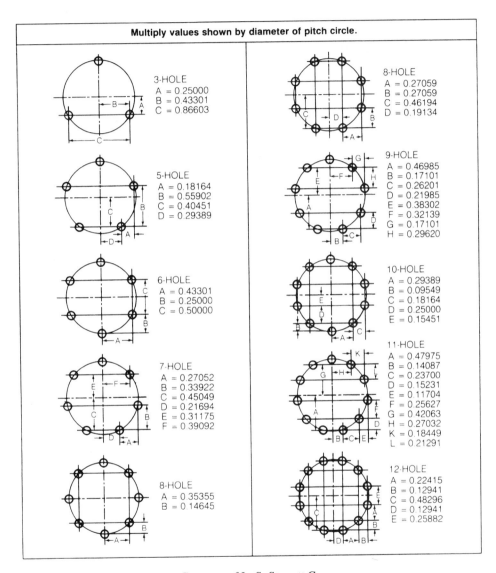

Courtesy of L. S. Starrett Co.

Notes:
1. Pitch circle, hole circle, and bolt circle mean the same thing.
2. This chart is very useful for conveniently converting bolt circle diameters to coordinates for inspection of a bolt circle on the surface plate.
3. Keep this chart handy for use as needed. It is recognized throughout the industry as an acceptable tool for use in the shop.

APPENDIX D   Differences between ASME and IOS Standard Symbols

Symbol	ASME Standard	ISO Standard
Least Material Condition	Ⓛ	None
Countersink	∨	None
Counterbore	⌴	None
Depth	↧	None
Spherical Radius	**SR**	None
Spherical Diameter	**S⌀**	None

The symbols shown are only those which are different in the ISO standard for geometric tolerancing. All others are the same as the ASME standard.

APPENDIX E Other Drawing Symbols

Symbol	ASME Standard	ISO Standard
All Around Profile	⌀ (circle on leader)	None
Dimension Origin	⊕	None
Conical Taper	▷—	▷—
Slope	◺	◺
Square Shape	□	□
Dimension Not to Scale	.200 (underlined)	.200 (underlined)
Number of Places	X	X
Arc Length	⌢.200	⌢.200
Radius	R	R
Controlled Radius	CR	None

APPENDIX F Table of Trigonometric Functions*

ANGLE	Sin	Cos	Tan	ANGLE	Sin	Cos	Tan
5°	.08716	.99619	.08749	45°30'	.71325	.70091	1.0176
5°30'	.09585	.99540	.09629	50°	.76604	.64279	1.1918
10°	.17365	.98481	.17633	50°30'	.77162	.63608	1.2131
10°30'	.18224	.98325	.18534	55°	.81915	.57358	1.4281
15°	.25882	.96593	.26795	55°30'	.82413	.56641	1.4550
15°30'	.26724	.96363	.27732	60°	.86603	.50000	1.7321
20°	.34202	.93969	.36397	60°30'	.87036	.49242	1.7675
20°30'	.35021	.93667	.37388	65°	.90631	.42262	2.1445
25°	.42262	.90631	.46631	65°30'	.90996	.41469	2.1943
25°30'	.43051	.90259	.47698	70°	.93969	.34202	2.7475
30°	.50000	.86603	.57735	70°30'	.94264	.33381	2.8239
30°30'	.50754	.86163	.58905	75°	.96593	.25882	3.7321
35°	.57358	.81915	.70021	75°30'	.96815	.25038	3.8667
35°30'	.58070	.81412	.71329	80°	.98481	.17365	5.6713
40°	.64279	.76604	.83910	80°30'	.98629	.16505	5.9758
40°30'	.64945	.76041	.85408	85°	.99619	.08716	11.430
45°	.70711	.70711	1.0000	85°30'	.99692	.07846	12.706

*This table is for example only. Refer to complete trigonometric tables (or scientific calculators) for practical purposes. This table was prepared using a scientific calculator.

Glossary

Angularity The condition for which a surface, axis, or centerplane must be inclined at a specific angle (other than 90°) with respect to a datum plane or datum axis.

Basic dimension A dimension that is specified on a drawing in a box. Basic dimensions are theoretical values used to describe the exact size, shape, or location of a feature. Basic dimensions have no tolerance. They are used to locate tolerance zones.

Centerplane A centerplane is the middle (or median plane) of a feature (particularly noncylindrical features such as slots, grooves, bosses, and tabs).

Circular runout A composite control of the circular elements of a surface that independently applies at any measured position while the part is being rotated 360° around a datum axis.

Circularity The condition of a surface of revolution (for example, cylinders or cones) for which all points of the surface intersected by any plane are equidistant from the center.

Coaxiality A term meaning two or more features (for example, a feature axis and a datum feature axis) have coincident axes. The coaxiality controls in the ASME Y14.5M standard are runout, concentricity, and position.

Concentricity A condition for which two or more features have a common axis. The lack of concentricity is *eccentricity*.

Cylindricity A condition of a surface of revolution for which all points of the surface are equidistant from a common axis.

Datum A theoretically exact point, axis, or plane that is derived from the true geometric counterpart of a specified datum feature. Datums are the origin from which many geometric tolerances are measured.

Datum feature A datum feature is an actual feature of a part that is used to establish a datum.

Datum reference frame A system of three mutually perpendicular datum planes or axes that is established from specified datum features.

Datum target A specified datum point, line, or area used to establish datum points, lines, planes, or areas for a special function, inspection repeatability.

Feature A general term meaning a physical characteristic of a part, such as a surface, hole, slot, pin, tab, or boss. Features of size (for example, holes, pins, or slots) are referred to as size features.

Feature control frame A rectangular box that contains the geometric tolerance symbol, the tolerance value (including modifiers on the tolerance value), and applicable datums in order of precedence (plus applicable modifiers on the datums).

Flatness The condition of a surface for which all elements are in one plane.

Form tolerances The allowable amount of variation in a feature from its desired form implied by the drawing. Form tolerances include flatness, straightness, circularity, and cylindricity.

Full indicator movement (FIM) The total movement observed on a dial indicator (or probe) during a measurement, such as parallelism or runout. Other terms, such as total indicator reading (TIR) and full indicator reading (FIR), have the same meaning.

Full indicator reading (FIR) See *full indicator movement*.

Intrinsic datum A term that applies when features are related to each other (for example, hole to hole or pin to pin).

Least material condition (LMC) A condition of a size feature when it contains the minimum amount of material allowed (for example, the largest hole or the smallest shaft). LMC is the opposite of MMC.

Location tolerance A tolerance that states how far an actual feature may vary from perfect location related to datums or other features. The location tolerances are position, concentricity, and symmetry.

Maximum material condition (MMC) The condition of a size feature when the feature contains the maximum amount of material allowed within size limits (for example, the largest shaft or the smallest hole).

Modifier A term often used to describe the application of MMC, LMC, or RFS principles. These modifiers, when used on a tolerance value, modify the stated tolerance.

Orientation tolerance Applies to related features when one feature is selected as the datum and the other feature is related to that datum. The orientation tolerances are parallelism, angularity, and perpendicularity.

Parallelism The condition of a surface, line, or axis that is equidistant at all points from a datum plane or axis.

Perpendicularity A condition for which a surface, axis, or line element is 90° from a datum plane or datum axis.

Position tolerance Formerly called *true position;* a position tolerance defines a zone within which the axis or centerplane of a feature is permitted to vary from true (perfect) position.

Profile of a line A tolerance that permits a uniform amount of profile variation (unilaterally or bilaterally) at each line element of a feature.

Profile of a surface A tolerance that permits a uniform amount of profile variation (unilaterally or bilaterally) on a surface.

Profile tolerance Profile tolerance controls the shape, size, and/or location.

Projected tolerance zone A tolerance zone that is applied to a hole in which a pin, stud, screw, bolt, or the like is to be inserted. A projected tolerance zone extends above the surface of a part to the functional length of the pin, stud, screw, or bolt relative to its assembly with the mating part.

Regardless of feature size (RFS) A condition for which the specified tolerance must be met regardless of the size of the feature being controlled.

Roundness See *Circularity*.

Runout A composite control of the desired form of a part surface of revolution during 360° rotation of the part on a datum axis. There are two types of runout tolerances in the ASME standard, circular runout and total runout.

Straightness A condition for which an element of a surface or an axis is in a straight line. There are two kinds of straightness tolerances in the ASME Y14.5M standard, straightness of surface elements and straightness of an axis (or a centerplane).

Symmetry A condition for which a feature (or features) is equally disposed about the centerplane of a datum feature.

Total indicator reading (TIR) See *Full indicator movement*.

Total runout A composite control of all elements of a surface at all circular and profile measuring positions as the part is rotated 360° about a datum axis.

True position A term used to describe the perfect (or exact) location of a point, line, or plane of a feature related to a datum reference.

Virtual condition The collective effect of size, form, and location error that must be considered when determining the fit between mating parts or features. The virtual condition represents the worst-case condition of assembly at MMC.

Solutions to Odd-Numbered Chapter Review Questions

CHAPTER 1

1. observer
3. zone
5. False (answer is *form*)
7. | // | .005 | A |
9. bonus
11. Ⓢ

CHAPTER 2

1. datums
3. True
5. True
7. | A–B |
9. highest

CHAPTER 3

1. form
3. LMC
5. True
7. True
9. ring

CHAPTER 4

1. False
3. .510"
5. wobble
7. d
9. monochromatic
11. optimum
13. c
15. True

CHAPTER 5

1. True
3. lines
5. two parallel planes
7. .756"
9. differential
11. Yes
13. True
15. True

CHAPTER 6

1. True
3. concentric
5. d
7. d
9. jets
11. c
13. d
15. False

CHAPTER 7

1. cylinders
3. runout
5. d
7. False
9. True

CHAPTER 8

1. True
3. True
5. c
7. False
9. True
11. d
13. boundary
15. True

CHAPTER 9

1. True
3. c
5. True
7. b
9. diameter symbol

CHAPTER 10

1. False
3. d
5. misalignment
7. False
9. d

CHAPTER 11

1. True
3. circular
5. True
7. a
9. True

CHAPTER 12

1. False
3. d
5. True
7. d
9. partial

CHAPTER 13

1. tangent
3. b
5. True
7. c
9. b

CHAPTER 14

1. True
3. three
5. False; not always
7. True
9. Yes

CHAPTER 15

1. d
3. total runout
5. d
7. d
9. c

CHAPTER 16

1. False
3. A
5. A
7. True
9. True
11. smaller
13. c
15. bonus

CHAPTER 17

1. datum
3. differential
5. two parallel
7. position
9. equal

CHAPTER 18

1. True
3. True
5. virtual
7. .4750″
9. True

Index

A

Accuracy and precision, 11
Additional tolerance, 244, 267, 289, 292, 313
Alternatives to hard gages, 32
Angularity, 146
    angular gage block method, 154
    definition, 147
    of a cone, 160
    secondary alignment datums, 152
    setting compound sine plates, 164
    sine-bar method, 149
    sine-plate method, 154
    of a size feature, 156
    surface to surface, 147
    tolerance zone, 147
    upside-down use of a sine bar, 165
ASME versus ISO symbols, 2

B

Basic dimensions, 3
Bolt circle chart, 339
Bonus tolerances, 6
Boundary gaging (size tolerances), 26, 29, 30

C

Circularity (roundness), 79
    bore gage method, 88
    effective size, 81
    lobes, 81
    pneumatic gaging method, 91
    polar graphs, 94
    precision spindle method, 91
    rotary table method, 89
    tolerance zone, 80
    vee-anvil micrometer method, 87
    vee-block method, 83
    verifying circularity using runout, 89
Circular runout, 168
    applications, 168
    applied to a cone, 176
    compound offset datum, 179
    end surface, 174
    inside diameter datum, 172
    vee-block method, 169
Coaxial controls, 244, 319, 320
Concentricity, 221
    definition, 221
    difference between concentricity and runout, 222
    differential measurement method, 223
    introduction, 221
    precision spindle method, 227
    verifying concentricity using total runout, 227
Contacting functional datums, 18
Conversion charts,
    bolt circle chart, 339
    coordinate measurements to cylindrical zones, 236, 337
    diametral zones to even ± coordinates, 236, 338

Coordinate dimensioning versus ASME Y14.5M, 232
Coordinate measuring machines, 293
Cylindricity, 96
    definition, 96
    inherent controls, 98
    open-setup methods, 98
    precision spindle methods, 100
    tolerance zone, 98
    verifying (using total runout), 98

D

Datums,
    compound datums, 19
    contacting datums, 18
    equalizing, 20
    functional, 14
    nonfunctional, 14, 21
    offset compound datums, 20
    qualified, 17
    reference frame, 15
    simulated, 16
    targets, 21
Diameter symbol, 2
Dimension (basic), 3
Dimension (reference), 3

E

Effective size (due to circularity error), 81
Equalizing datums, 20

F

Families of tolerances, 3
Feature control frame, 16
Flatness, 36
    definition, 36
    direct contact method, 43
    fixed-plane method, 41
    Indian pins methods, 43
    jackscrew method, 38
    optical flats method, 45
        contact method, 46
        wedge method, 46
    tolerance zone, 37
    wobble-plate method, 41
Free state variation, 5, 10

Fringes (interference light bands), 45
Full indicator movement (FIM), 3
Functional datums, 14
Functional gaging, 304
    advantages, 305
    alternatives, 32, 322
    costs, 306
    design principles, 304
    disadvantages, 306
    gage makers' tolerance, 309
    materials, 307
    perpendicularity, 315
    pins, 314
    position tolerance at MMC, 316-322
    straightness of an axis at MMC, 316
    tolerances that cannot be gaged, 310
    use of gages, 324
    virtual condition rule, 312
    wear allowance, 309

G

General rules, 7
Glossary, 343
Graphical inspection analysis (GIA), 253, 263

I

Implied 90 degree angles, 124
Indian pins, 43
Interference light bands, 45
Interpreting light bands, 47
Intrinsic datum, 258, 321
Inspecting size tolerances, 24

L

Least material condition, 5
Lobes of circularity error, 81

M

Maximum boundary of perfect form at MMC, 8
Maximum material condition (MMC), 5
Modifiers, 5

N

Nonfunctional datums, 21

O

Offset compound datums, 118
Optimum line, 54
Optimum plane, 36

P

Parallelism, 103
    axis to axis, 112
    axis to surface, 111
    bonus tolerance at MMC, 111, 237
    compound datums, 118
    definition, 104
    secondary alignment datums, 116
    surface plate method, 108
    surface to surface, 105
    tolerance zone, 104
Perpendicularity, 123
    axis to axis (RFS), 140
    axis to surface (MMC), 139, 315
    axis to surface (RFS), 135
    cylindrical square method, 126
    definition, 124
    functional gaging at MMC, 315
    implied perpendicularity, 124
    precision square method, 126
    right angle plate method, 129
    secondary alignment datums, 130
    single surface (one datum), 124
    surface to surface, 124
    tolerance zone, 124
Pitch diameter rule, 9
Position tolerances, 231
    additional tolerance, 244, 267, 289, 292, 313, 326
    best fit, 248, 321
    bi-directional tolerance, 274
    boundary, 287
    coaxial features (MMC), 244, 319, 320
    composite tolerance, 271
    conical zone, 290
    coordinate measuring machines (CMM), 293
    functional gaging at MMC, 316, 319, 320, 321, 322, 326
    introduction, 231
    paper gaging, 253, 263,
    pattern locating tolerances, 247
        cylindrical parts, tertiary datum, 267
    intrinsic datum, 258, 321
    LMC application, 284, 291
    One single segment, 323
    perpendicularity control, 263, 313
    primary datum, 263, 313
    projected tolerance zone, 5, 279
    separate requirements, 288
    size feature datum, 313
    single feature location, 238, 316
    tertiary datum, 265, 267
    two single segment, 285
    virtual condition datum, 313
    zero tolerance at MMC, 282, 324
Precision spindle, 91
Profile of a line, 193
    datum specified, 201
    general measurement, 194
    introduction, 193
    mechanical gaging, 203
    no datum specified, 195
    optical comparator method, 197
    tolerance zones, 194, 195
Profile of a surface, 207
    coplanar surfaces, 218
    datums specified, 210
    introduction, 207
    jackscrew method (coplanar), 218
    no datum specified, 208
    tolerance zone, 208
Profile tolerances, 193, 207
Projected tolerance zone, 278

Q

Qualified datums, 17

R

Reference dimensions, 3
Regardless of feature size (RFS), 5
Restrained features, 11
Roundness (see Circularity)
Rules (general), 7
Rule 1, 7
Rule 2, 8
Rule (pitch diameter), 9
Rule (datum/virtual condition), 9
Runout tolerance (circular), 168
Runout tolerance (total), 183

S

Secondary alignment datum, 130, 150, 152
Simulated datum, 16
Simulating boundary gaging, 26, 29, 30
Sine bars, 149
Size tolerances, 24
Squareness (See Perpendicularity)
Straightness tolerances, 50
Straightness of a center plane, 53
Straightness of an axis, 53, 66
    differential measurement, 66
    functional gage (at MMC), 74, 317
    introduction, 50
    precision spindle method, 68
    tolerance zone, 52, 53
Straightness of surface elements, 52
    introduction, 53
    jack-screw method, 53
    non-cylindrical parts, 59
    optical comparator method, 58
    optimum line, 54
    precision straight edge method, 57
    tolerance zone, 52, 59
    two-block method, 58
Symbols, 2, 340, 341
Symmetry, 299

T

Target datums, 21
Tangent Plane, 5
Tolerance (additional), 244, 267, 289, 292, 313
Tolerances of size, 24
Tolerance zones, 4
Total indicator reading (TIR), 3
Total runout, 183
    end surface application, 189
    inside diameter datum, 186
    introduction, 183
    outside diameter datums, 184
    vee-block method, 186
Trigonometric functions table (example), 342

V

Virtual condition, 5
Virtual datum, 9

W

Wear allowance (functional gages), 309
Wobble plate method (flatness), 41

Z

Zones (tolerance), 4